Laboratory Animal Management

DOGS

Committee on Dogs
Institute of Laboratory Animal Resources
Commission on Life Sciences
National Research Council

NATIONAL ACADEMY PRESS
Washington, D.C.   1994

National Academy Press • 2101 Constitution Avenue, N.W. • Washington, D.C. 20418

NOTICE: The project that is the subject of this report was approved by the Governing Board of the National Research Council, whose members are drawn from the councils of the National Academy of Sciences, National Academy of Engineering, and Institute of Medicine. The members of the committee responsible for the report were chosen for their special competencies and with regard for appropriate balance.

This report has been reviewed by a group other than the authors according to procedures approved by a Report Review Committee consisting of members of the National Academy of Sciences, National Academy of Engineering, and Institute of Medicine. ·

This study was supported by the U.S. Department of Health and Human Services (DHHS) through contract number NO1-CM-07316 with the Division of Cancer Treatment, National Cancer Institute; the Animal Welfare Information Center, National Agricultural Library, U.S. Department of Agriculture (USDA), through grant number 59-32U4-8-60; and Regulatory Enforcement and Animal Care, Animal and Plant Health Inspection Service, USDA, through grant number 59-32U4-8-60. Additional support was provided by the following members of the Pharmaceutical Manufacturers Association: Berlex Laboratories, Inc., Cedar Knolls, New Jersey; Bristol-Myers Squibb Co., New York, New York; Bristol-Myers Research, Princeton, New Jersey; Burroughs Wellcome Co., Research Triangle Park, North Carolina; Dupont Merck Research & Development, Wilmington, Delaware; Johnson & Johnson, New Brunswick, New Jersey; Marion Merrell Dow Inc., Kansas City, Missouri; Pfizer Inc., Groton, Connecticut; Schering-Plough Research, Bloomfield, New Jersey; SmithKline Beecham Pharmaceuticals, Swedeland, Pennsylvania; and Syntex Research, Palo Alto, California.

ILAR's core program is supported by grants from the National Center for Research Resources, National Institutes of Health; National Science Foundation; American Cancer Society, Inc.; and U.S. Army Medical Research and Development Command, which is the lead agency for combined Department of Defense funding also received from the Human Systems Division, Air Force Systems Command; Armed Forces Radiobiology Research Institute; Uniformed Services University of the Health Sciences; and U.S. Naval Medical Research and Development Command.

Any opinions, findings, and conclusions or recommendations expressed in this publication are those of the committee and do not necessarily reflect the views of DHHS, USDA, or other sponsors, nor does the mention of trade names, commercial products, or organizations imply endorsement by the U.S. government or other sponsor.

Library of Congress Cataloging-in-Publication Data

Dogs : laboratory animal management / Committee on Dogs, Institute of
    Laboratory Animal Resources, Commission on Life Sciences, National
    Research Council.
       p.   cm.
     Includes bibliographical references and index.
     ISBN 0-309-04744-7
     1. Dogs as laboratory animals. I. Institute of Laboratory Animal
    Resources (U.S.). Committee on Dogs.
    SF407.D6D64   1994
    636.7′0885—dc20   94-960
                                                CIP

Printed in the United States of America

COMMITTEE ON DOGS

Fred W. Quimby (*Chairman*), Center for Research Animal Resources, New York State College of Veterinary Medicine, Cornell University, Ithaca, New York

Emerson L. Besch, Department of Physiological Sciences, University of Florida College of Veterinary Medicine, Gainesville, Florida

Linda C. Cork, Department of Comparative Medicine, Stanford University, Stanford, California

Suzanne Hetts, Humane Society of Denver, Denver, Colorado

Warren C. Ladiges, Department of Comparative Medicine, University of Washington, Seattle, Washington

Richard J. Traystman, Department of Anesthesiology and Critical Care Medicine, The Johns Hopkins Hospital, Baltimore, Maryland

Staff

Dorothy D. Greenhouse, Project Director
Amanda E. Hull, Project Assistant
Norman Grossblatt, Editor

The Institute of Laboratory Animal Resources (ILAR) was founded in 1952 under the auspices of the National Research Council. A component of the Commission on Life Sciences, ILAR serves as a coordinating agency and a national and international resource for compiling and disseminating information on laboratory animals, promoting education, planning and conducting conferences and symposia, surveying existing and required facilities and resources, upgrading laboratory animal resources, and promoting high-quality, humane care of laboratory animals in the United States.

CONTRIBUTORS

Gregory M. Acland, James A. Baker Institute, Cornell University, Ithaca, New York

Judith A. Bell, Marshall Research Animals, North Rose, New York

Dwight D. Bowman, New York State College of Veterinary Medicine, Cornell University, Ithaca, New York

David P. Brooks, SmithKline Beecham Pharmaceuticals, King of Prussia, Pennsylvania

Phillip R. Brown, Division of Comparative Medicine, The Johns Hopkins University School of Medicine, Baltimore, Maryland

Robert W. Bull, Michigan State University, East Lansing, Michigan

Leland E. Carmichael, James A. Baker Institute, Cornell University, Ithaca, New York

J. Derrell Clark, Animal Resources, University of Georgia College of Veterinary Medicine, Athens, Georgia

Patrick W. Concannon, New York State College of Veterinary Medicine, Cornell University, Ithaca, New York

Lawrence G. Carbone, New York State College of Veterinary Medicine, Cornell University, Ithaca, New York

Laurel J. Dungan, Department of Comparative Medicine, Medical University of South Carolina, Charleston, South Carolina

W. Jean Dodds, Hemopet, Santa Monica, California

Robin D. Gleed, New York State College of Veterinary Medicine, Cornell University, Ithaca, New York

Arthur S. Hall, Department of Animal Care, Oregon Health Sciences University, Portland, Oregon

Margaret S. Landi, SmithKline Beecham Pharmaceuticals, King of Prussia, Pennsylvania

George Lust, James A. Baker Institute, Cornell University, Ithaca, New York

Ronald R. Minor, New York State College of Veterinary Medicine, Cornell University, Ithaca, New York

Bruce A. Muggenburg, Inhalation Toxicology Research Institute, Albuquerque, New Mexico

Bryan E. Ogden, Department of Animal Care, Oregon Health Sciences University, Portland, Oregon

Donald F. Patterson, Section of Medical Genetics, University of Pennsylvania School of Veterinary Medicine, Philadelphia, Pennsylvania

Arleigh Reynolds, New York State College of Veterinary Medicine, Cornell University, Ithaca, New York

Robert M. Shull, Department of Pathobiology, University of Tennessee College of Veterinary Medicine, Knoxville, Tennessee

Alison C. Smith, Department of Comparative Medicine, Medical University of South Carolina, Charleston, South Carolina

Rainer F. Storb, Fred Hutchinson Cancer Research Center, Seattle, Washington

M. Michael Swindle, Department of Comparative Medicine, Medical University of South Carolina, Charleston, South Carolina

Beth A. Valentine, Department of Pathology, New York State College of Veterinary Medicine, Cornell University, Ithaca, New York

David A. Valerio, Hazleton Research Products, Denver, Pennsylvania

The National Academy of Sciences is a private, nonprofit, self-perpetuating society of distinguished scholars engaged in scientific and engineering research, dedicated to the furtherance of science and technology and to their use for the general welfare. Upon the authority of the charter granted to it by the Congress in 1863, the Academy has a mandate that requires it to advise the federal government on scientific and technical matters. Dr. Bruce M. Alberts is president of the National Academy of Sciences.

The National Academy of Engineering was established in 1964, under the charter of the National Academy of Sciences, as a parallel organization of outstanding engineers. It is autonomous in its administration and in the selection of its members, sharing with the National Academy of Sciences the responsibility for advising the federal government. The National Academy of Engineering also sponsors engineering programs aimed at meeting national needs, encourages education and research, and recognizes the superior achievements of engineers. Dr. Robert M. White is president of the National Academy of Engineering.

The Institute of Medicine was established in 1970 by the National Academy of Sciences to secure the services of eminent members of appropriate professions in the examination of policy matters pertaining to the health of the public. The Institute acts under the responsibility given to the National Academy of Sciences by its congressional charter to be an adviser to the federal government and upon its own initiative to identify issues of medical care, research, and education. Dr. Kenneth I. Shine is president of the Institute of Medicine.

The National Research Council was established by the National Academy of Sciences in 1916 to associate the broad community of science and technology with the Academy's purposes of furthering knowledge and advising the federal government. Functioning in accordance with general policies determined by the Academy, the Council has become the principal operating agency of both the National Academy of Sciences and National Academy of Engineering in the conduct of their services to the government, the public, and the scientific and engineering communities. The Council is administered jointly by both Academies and the Institute of Medicine. Dr. Bruce M. Alberts and Dr. Robert M. White are chairman and vice-chairman, respectively, of the National Research Council.

Preface

It has been 2 decades since the Institute of Laboratory Animal Resources first published *Dogs: Standards and Guidelines for the Breeding, Care, and Management of Laboratory Animals* (National Academy of Sciences, Washington, D.C., 1973). During that period, great strides have been made in improving care and management techniques, making available specific-pathogen-free and purpose-bred dogs, and identifying dogs with precisely defined genetic disorders. The dog has proved to be "man's best friend," not only because it is considered a companion and family member, but also because its use in research has been associated with many breakthrough discoveries in human medicine (e.g., the discovery of insulin as a treatment for type I diabetes mellitus).

The same period has been characterized by increased public awareness and scrutiny of research funding, occupational health and safety, and animal welfare. New federal and state laws specifically intended to protect research animals have been promulgated and regulations established. In addition to presenting information relevant to the care and use of dogs in research and making recommendations based on an objective evaluation of that information, it was the committee's intent to incorporate in this report those aspects of canine husbandry embodied in federal law. Federal regulations and policies protecting dogs in research are therefore summarized in Chapter 1, which provides information for obtaining copies. Specific details of the regulations and policies are given throughout the text.

The committee firmly believes that good research requires a good animal-care program. The committee is also aware of the tremendous variation in physiologic traits among canine models. Dogs vary greatly in size, age, health status, physical conformation of the breed, behavioral characteristics, and experience. Therefore, no standard of animal care is likely to be optimal for all dogs. The committee recommends that performance standards be used with sound professional judgment in implementing the animal-care program.

Readers who detect errors of omission or commission or who have evidence to support improved procedures are invited to send suggestions to ILAR, National Research Council, 2101 Constitution Avenue, Washington, DC 20418.

The committee wishes to thank the entire staff of ILAR, but especially Dr. Dorothy Greenhouse and Ms. Amanda Hull, for assisting in the production of this manuscript. The committee also acknowledges the many fine contributions made to this report by scientists specializing in the care and use of dogs in research; their names appear on pages *iv* and *v*.

Fred W. Quimby, *Chairman*
Committee on Dogs

Contents

1 INTRODUCTION   1
   References   3

2 CRITERIA FOR SELECTING EXPERIMENTAL ANIMALS   4
   Genetic Factors   5
   Biologic Factors   7
   Behavioral Factors   7
   Hazards   9
   References   9

3 HUSBANDRY   11
   Housing   12
   Exercise and Environmental Enrichment   21
   Food   24
   Water   26
   Bedding and Resting Apparatuses   26
   Sanitation   27
   Identification and Records   27
   Emergency, Weekend, and Holiday Care   29
   Transportation   39
   References   32

4   MANAGEMENT OF BREEDING COLONIES   35
    Reproduction   35
    Neonatal Care   40
    Reproductive Problems   41
    Special Nutritional Requirements   42
    Vaccination and Deworming   43
    Socialization of Pups   44
    Record Keeping   46
    References   47

5   VETERINARY CARE   51
    Procurement   52
    Control of Infectious Diseases   53
    Control of Parasitic Diseases   57
    Recognition and Alleviation of Pain and Distress   63
    Surgery and Postsurgical Care   68
    Euthanasia   70
    References   72

6   SPECIAL CONSIDERATIONS   76
    Protocol Review   76
    Restraint   78
    Special Care for Animal Models   78
        Aging   79
        Cardiovascular Diseases   81
        Ehlers-Danlos Syndrome   91
        Endocrinologic Diseases   93
        Hematologic Disorders   97
        Immunologic Diseases   101
        Lysosomal Storage Diseases   107
        Muscular Dystrophy   110
        Neurologic Disorders   112
        Ophthalmologic Disorders   114
        Orthopedic Disorders   116
        Radiation Injury   117
        Gene Therapy   119
    References   122

APPENDIX: CROSS REFERENCE   131

INDEX   133

Laboratory Animal Management

DOGS

1

Introduction

Dogs make valuable contributions in biomedical research because they share many biochemical and physiologic characteristics with humans and spontaneously develop disorders that are homologous to pathologic conditions in humans. While using them as models for human disease, we have also learned much about normal physiologic processes in dogs themselves. Advances in molecular genetics, reproduction, behavior, immunology, hematology, endocrinology, microbiology, nutrition, pharmacology, and oncology, to name a few, have made dogs more valuable as models and, at the same time, have provided veterinarians with useful information for the diagnosis and treatment of canine diseases.

In the past 2 decades, two amendments to the Animal Welfare Act (in 1976 and 1985) and a section added to the Health Research Extension Act of 1985 have resulted in revised standards for dogs. Institutions that use dogs must comply with the Code of Federal Regulations, Title 9, Subchapter A, Parts 1-3 (9 CFR 1-3), commonly called the Animal Welfare Regulations (AWRs), which were promulgated to administer the Animal Welfare Act. Institutions receiving Public Health Service (PHS) funding must also comply with the *Public Health Service Policy on Humane Care and Use of Laboratory Animals* (hereafter called the *PHS Policy*) (PHS, 1986), which in turn requires compliance with the *Guide for the Care and Use of Laboratory Animals* (hereafter called the *Guide*) (NRC, 1985). Some of the AWRs are based on engineering standards (e.g., that on space requirements for dogs), but most rely on performance standards (i.e., the demonstration of animal well-being). It is expected, therefore, that professional judgment will be used in applying the AWRs. It is also incumbent on all people using dogs to seek improvements in the methods for using them.

This edition of *Dogs: Laboratory Animal Management* incorporates features of housing, management, and care that are related to the expanded use of dogs as models of human diseases and an intrepretative summary of the AWRs and requirements of the *PHS Policy.* The appendix lists subjects within this text by page number with cross references to corresponding sections in the AWRs and the *Guide.* The regulations, policies, and guidelines that are applicable to dogs include the following:

•   Code of Federal Regulations, Title 9, Subchapter A, Parts 1-3 (commonly called the Animal Welfare Regulations). Available from Regulatory Enforcement and Animal Care, APHIS, USDA, Federal Building, Room 565, 6505 Belcrest Road, Hyattsville, MD 20782 (telephone, 301-436-7833).

•   *Public Health Service Policy on Humane Care and Use of Laboratory Animals.* Available in English or Spanish from the Office for Protection from Research Risks, Building 31, Room 5B59, NIH, Bethesda, MD 20892 (telephone, 301-496-7163).

•   *Guide for the Care and Use of Laboratory Animals.* Available in English or Spanish from the Office for Protection from Research Risks, Building 31, Room 5B59, NIH, Bethesda, MD 20892 (telephone: 301-496-7163). Single copies (English only) available from the Institute of Laboratory Animal Resources, National Research Council, 2101 Constitution Avenue NW, Washington, DC 20418 (telephone, 202-334-2590).

•   Code of Federal Regulations, Title 21, Part 58; Title 40, Part 160; and Title 40, Part 792 (commonly called the Good Laboratory Practice, or GLP, Standards). Available from the Superintendent of Documents, U.S. Government Printing Office, Washington, DC 20402 (telephone, 202-783-3238).

•   IATA Live Animal Regulations. Available in English, French, or Spanish from the International Air Transport Association (IATA), 2000 Peel Street, Montreal, Quebec, Canada H3A 2R4 (telephone, 514-844-6311).

All animals used in research must be treated with the dignity and respect due living beings. Those who use animals in experiments must therefore be properly trained in methods appropriate for the species used. It is the responsibility of each research facility to develop educational programs for animal-care providers and the research staff (9 CFR 2.32). Recommendations for establishing such programs have recently been published (NRC, 1991).

REFERENCES

NRC (National Research Council), Institute of Laboratory Animal Resources, Committee on Care and Use of Laboratory Animals. 1985. Guide for the Care and Use of Laboratory Animals. NIH Pub. No. 86-23. Washington, D.C.: U.S. Department of Health and Human Services. 83 pp.

NRC (National Research Council), Institute of Laboratory Animal Resources, Committee on Educational Programs in Laboratory Animal Science. 1991. Education and Training in the Care and Use of Laboratory Animals: A Guide for Developing Institutional Programs. Washington, D.C.: National Academy Press. 139 pp.

PHS (Public Health Service). 1986. Public Health Service Policy on Humane Care and Use of Laboratory Animals. Washington, D.C.: U.S. Department of Health and Human Services. 28 pp.

2

Criteria for Selecting Experimental Animals

Scientists who are planning experiments evaluate both animal and nonanimal approaches. If there are no suitable alternatives to the use of live animals, the appropriate species is selected on the basis of various scientific and practical factors, including the following:

- Which species will yield the most scientifically accurate and interpretable results?
- According to critical review of the scientific literature, which species have provided the best, most applicable historical data?
- On which species will data from the proposed experiments be most relevant and useful to present and future investigators?
- Which species have special biologic or behavioral characteristics that make them most suitable for the planned studies?
- Which species have features that render them inappropriate for the planned studies?
- Which species present the fewest or least severe biologic hazards to the research team?
- Which species require the fewest number of animals?
- Which species that meet the above criteria are most economical to acquire and house?

For many scientific experiments, the answer to those questions will be the domestic dog, *Canis familiaris*. The size, biologic features, and cooperative, docile nature of the well-socialized dog make it the model of choice for a variety of scientific inquiries. The contributions of the dog to human health and well-being are numerous (Gay, 1984).

Although research with dogs is often primarily to benefit humans, it has also greatly benefited dogs that are kept as companion animals. Examples of the benefits to dogs are improvements in diagnostic techniques; treatments for diabetes and arthritis; surgical procedures for correcting or treating cardiovascular, orthopedic, and neurologic disorders; and therapies for bacterial, neoplastic, and autoimmune diseases. Moreover, dogs have been necessary for the development of vaccines that protect companion animals against viral diseases (e.g., distemper and parvovirus disease) and drugs that prevent parasitic diseases (e.g., dirofilariasis, or heartworm disease).

GENETIC FACTORS

All domestic dogs, irrespective of breed, are *Canis familiaris*. Canine genotypes and phenotypes vary among breeds as a result of selective breeding, which has created variations in allele frequency between breeds. Although "pure" breeds might have a higher frequency of some genes, much genetic variation remains in most breeds.

The canine karyotype consists of 78 chromosomes (Minouchi, 1928). Most of the autosomes are acrocentric or telocentric, and many pairs do not differ markedly in size. Recently, an improved method for staining canine chromosomes has been developed that makes karyotyping with Giemsa banding feasible (Stone et al., 1991).

A number of loci have been identified that code for the antigens of the canine major histocompatibility complex, which has been designated DLA (Vriesendorp et al., 1977). Initially, several alleles were defined with serologic techniques at three class I loci, and several alleles were defined with cellular techniques at a DLA class II locus (Bull et al., 1987; Deeg et al., 1986). Molecular techniques are being used to refine the definition of the DLA class I loci, and at least eight class I genes have been demonstrated in the dog (Sarmiento and Storb, 1989). Molecular-genetic studies to characterize canine class II loci correlate well with earlier work in which techniques for cell typing for class II antigens were used (Sarmiento and Storb, 1988a,b). The characterization of canine DLA loci is extremely useful for transplantation studies (Ladiges et al., 1985) and for demonstrating an association between the major histocompatibility complex and some inherited canine diseases (Teichner et al., 1990).

Attempts are under way to develop maps that identify the location of canine genes that control particular traits (e.g., inherited diseases and such behavioral tendencies as herding and aggression). Two approaches are used. The first relies on the principle that the relative positions of genes in a particular region of DNA are comparable in humans, dogs, and other species. Conserved regions can be identified in DNA samples with restriction-fragment length polymorphisms (usually called RFLPs) that have been identified with probes for human and murine genes whose chromosomal locations are known. To enhance the detection of polymorphisms, investigators sometimes produce dog-coyote hybrids, cross-breed two widely divergent dog breeds, or analyze a large, well-defined canine kindred (Joe Templeton, Department of Veterinary Pathobiology, College of Veterinary Medicine, Texas A&M University, College Station, Tex., personal communication, 1993). The second approach uses simple sequence-repeat polymorphisms (microsatellite probes). Specific simple sequence-repeat markers that are highly polymorphic in dogs have been developed to study the canine genome (Ostrander et al., 1992, 1993). These and other techniques, such as chromosomal in situ hybridization and somatic cell hybridization, will likely greatly increase our understanding of canine genetics.

Inherited defects—including lysosomal storage diseases, retinal degenerations, coagulopathies, complement deficiency, and various musculoskeletal, hematopoietic, immunologic, and neurologic diseases—are common in purebred dogs, and many specific disorders are found most commonly in particular breeds (Patterson et al., 1989). This phenomenon might be related, in part, to breeders' inadvertent selection for mutant alleles that are closely linked to loci that determine breed-typical traits or to the chance increase in frequency of particular mutant alleles caused by the founder effect or random genetic drift. The high frequency of inherited canine disorders (compared with murine disorders) was recognized as early as 1969 (Cornelius, 1969). During the 20-year period 1960-1980, 20 percent of more than 1,200 literature citations on naturally occurring animal models of human diseases involved dogs (Hegreberg and Leathers, 1980). A compilation in 1989 noted that 281 inherited disease entities had been reported in dogs (Patterson et al., 1989). Many of those constitute the only animal models for investigating the corresponding human diseases (Patterson et al., 1988). The 19-fascicle *Handbook: Animal Models of Human Disease* (RCP, 1972-1993) lists 83 canine models of human diseases, many of which are hereditary, and the two-volume *Spontaneous Animal Models of Human Disease* (Andrews et al., 1979) describes many canine models.

In scientific studies in which genetic uniformity is desirable or in long-term studies in which the expected differences between experimental and control subjects are likely to be small, purpose-bred dogs (e.g., beagles) might be a more appropriate choice than dogs of unknown provenance. An advantage of using beagles, as opposed to other purpose-bred dogs, is the potential availability of other members of the kindred. But if the studies are to determine the greatest range of a variable that is likely to occur among the experimental subjects or if the experiments are of short duration, random-source dogs might be more useful and less expensive (see "Procurement" in Chapter 5).

BIOLOGIC FACTORS

Dogs are monogastric carnivores with a short generation time (i.e., the calculated interval between when a pup is born and when its first offspring could be born) and a maximum life span of approximately 20 years; larger breeds appear to have a shorter maximum life span than smaller breeds. The canine mortality rate doubles every 3 years, compared with every 0.3 year for the rat (maximum life span, 5.5 years), every 15 years for the rhesus monkey (maximum life span, more than 35 years), and every 8 years for humans (maximum life span, more than 110 years) (Finch et al., 1990). Dogs are useful models for studying the lifetime effects of environmental factors, and there is an extensive literature on their use in radiation biology (see Gay, 1984; Shifrine and Wilson, 1980).

Selective breeding has resulted in a spectrum of behaviors and a large range of canine body sizes, from the giant breeds (e.g., Irish wolfhound), which can measure 91 cm (36 in) at the shoulder and weigh more than 56 kg (124 lb), to the toy breeds (e.g., Pomeranian), which can measure less than 31 cm (12 in) in height and weigh less than 4.5 kg (10 lb). Larger dogs, which can include mongrels or dogs of unknown breeding, are particularly well suited to cardiovascular, transplantation, and orthopedic studies, because body weights and blood volumes approximate those of humans (see Gay, 1984; Shifrine and Wilson, 1980; Swindle and Adams, 1988). The dog's size also lends itself to procedures that cannot be carried out in smaller species, e.g., when the instrumentation essential for collecting scientific data is bulky and cannot be miniaturized and when the resolution of imaging equipment requires a larger target field than is available in a small animal.

An individual dog often can be studied in great detail or in many ways, which might reduce the number of subjects needed for a study and generate a more definitive data set. For example, it is possible to take multiple blood samples of several milliliters each from a single dog over some period without compromising the dog's well-being, but taking samples of similar size during the same period from a single mouse or rat would be impossible.

BEHAVIORAL FACTORS

The social unit for dogs is the pack, and most dogs can be socialized to accept humans as the dominant individual in their social hierarchy, especially if the techniques used to socialize them provide rewarding experiences (e.g., food treats, petting, and verbal reinforcements) and minimize

TABLE 2.1   Selected Canine[a] Zoonoses

Disease in Humans	Agent	Mode of Transmission (Intermediate Host or Vector)[b]
Acariasis	*Cheyletiella yasguri*	Direct
Amebiasis	*Entamoeba histolytica*	Direct
American trypanosomiasis (Chagas' disease)	*Trypanosoma cruzi*	Indirect (triatomine insect)
Brucellosis	*Brucella canis*	Direct
Campylobacteriosis	*Campylobacter jejuni*	Direct
Coenurosis	*Taenia multiceps*	Direct
Colibacillosis	Enteropathogenic *Escherichia coli*	Direct
Cutaneous larva migrans	*Ancylostoma braziliense* *Ancylostoma caninum*	Direct
Dipylidiasis	*Dipylidium caninum*	Indirect (dog flea)
Df2 infections	Dysgonic fermenter-2	Direct
Dirofilariasis	*Dirofilaria immitis* *Dirofilaria repens*	Indirect (mosquito)
Giardiasis	*Giardia intestinalis* (*canis*)	Direct
Hydatidosis	*Echinococcus granulosus*	Direct
Larva currens	*Strongyloides stercoralis*	Direct
Leishmaniasis (cutaneous)	*Leishmania braziliensis peruviana*	Indirect (phlebotomine flies)
Leishmaniasis (visceral)	*Leishmania donovani*	Indirect (phlebotomine flies)
Leptospirosis	*Leptospira* spp. (usually *L. canicola*)	Direct
Pasteurellosis	*Pasteurella multocida*	Direct
Rabies	Rabies virus	Direct
Ringworm (dermatomycoses)	*Microsporum canis* *Trichophyton mentagrophytes*	Direct
Rocky Mountain spotted fever	*Rickettsia rickettsii*	Indirect (tick)
Salmonellosis	*Salmonella* spp.	Direct
Scabies	*Sarcoptes scabiei*	Direct
Tularemia	*Francisella tularensis*	Indirect (tick)
Visceral larva migrans	*Toxacara canis* *Toxascaris leonina*	Direct
Yersiniosis	*Yersinia enterocolitica*	Direct

[a]North, Central, and South American dogs.

[b]Direct = transmission by direct contact with the dog, its excretions, or its secretions; no other vector or intermediate host is required.

aversive experiences.   Different breeds and individual dogs differ in the ease and rapidity with which they can be socialized to humans (Scott and Fuller, 1965).   However, properly socialized dogs can be docile and can be trained to cooperate in procedures that require repeated contacts with research personnel.   For example, most dogs will allow venipuncture with minimal restraint and will cooperate during detailed physical and neurologic evaluations.

HAZARDS

Unvaccinated dogs might harbor rabies virus, and preexposure immunization should be made available to personnel who are at substantial risk of infection (NRC, 1985). Dogs also have internal and external parasites that can be shared with humans (see "Parasitic Diseases" in Chapter 5). Table 2.1 lists selected zoonoses, zoonotic agents, and modes of transmission. Detailed discussions of zoonoses have been published (Acha and Szyfres, 1987; August, 1988; Elliot et al., 1985; Fishbein and Robinson, 1993; Hubbert et al., 1975). Personnel can develop allergies to canine dander and saliva, can be bitten or scratched, might suffer hearing impairment from prolonged exposure to excessive noise generated by barking dogs or mechanical equipment, or can be injured while lifting or transporting large dogs. To deal with these and other animal-related health problems, institutions must provide occupational health programs for personnel who work in animal facilities or have substantial animal contact (NRC, 1985).

REFERENCES

Acha, P. N., and B. Szyfres. 1987. Zoonoses and Communicable Diseases Common to Man and Animals, 2d ed. Scientific Pub. No. 503. Washington, D.C.: Pan American Health Organization. 963 pp.

Andrews, E. J., B. C. Ward, and N. H. Altman, eds. 1979. Spontaneous Animal Models of Human Disease. New York: Academic Press. Vol. I, 322 pp.; vol. II, 324 pp.

August, J. R. 1988. Dygonic fermenter-2 infections. J. Am. Vet. Med. Assoc. 193:1506-1508.

Bull, R. W., H. M. Vriesendorp, R. Cech, H. Grosse-Wilde, A. M. Bijma, W. L. Ladiges, K. Krumbacher, I. Doxiadis, H. Ejima, J. Templeton, E. D. Albert, R. Storb, and H. J. Deeg. 1987. Joint report of the Third International Workshop on Canine Immunogenetics. II. Analysis of the serological typing of cells. Transplantation 43:154-161.

Cornelius, C. E. 1969. Animal models—A neglected medical resource. N. Engl. J. Med. 281:934-944.

Deeg, H. J., R. F. Raff, H. Grosse-Wilde, A. M. Bijma, W. A. Buurman, I. Doxiadis, H. J. Kolb, K. Krumbacher, W. Ladiges, K. L. Losslein, G. Schoch, D. L. Westbroek, R. W. Bull, and R. Storb. 1986. Joint report of the Third International Workshop on Canine Immunogenetics. I. Analysis of homozygous typing cells. Transplantation 41:111-117.

Elliot, D. L., S. W. Tolle, L. Goldberg, and J. B. Miller. 1985. Pet-associated illness. N. Engl. J. Med. 313:985-995.

Finch, C. E., M. C. Pike, and M. Witten. 1990. Slow mortality rate accelerations during aging in some animals approximate that of humans. Science 249:902-905.

Fishbein, D. B., and L. E. Robinson. 1993. Rabies. N. Engl. J. Med. 329:1632-1638.

Gay, W. I. 1984. The dog as a research subject. The Physiologist 27:133-141.

Hegreberg, G., and C. Leathers, eds. 1980. Bibliography of Naturally Occurring Animal Models of Human Disease. Pullman, Washington: Student Book Corp. 146 pp.

Hubbert, W. T., McCulloch, W. F., and Schnurrenberger, P. R., eds. 1975. Diseases Transmitted from Animals to Man, 6th ed. Springfield: Ill: Charles C Thomas. 1,236 pp.

Ladiges, W. C., H. J. Deeg, R. F. Raff, and R. Storb. 1985. Immunogenetic aspects of a canine breeding colony. Lab. Anim. Sci. 35(1):58-62.

Minouchi, O. 1928. The spermatogenesis of the dog with special reference to meiosis. Jpn. J. Zool. 1:255-268.

NRC (National Research Council), Institute of Laboratory Animal Resources, Committee on Care and Use of Laboratory Animals. 1985. Guide for the Care and Use of Laboratory Animals. NIH Pub. No. 86-23. Washington, D.C.: U.S. Department of Health and Human Services. 83 pp.

Ostrander, E. A., P. M. Jong, J. Rine, and G. Duyk. 1992. Construction of small-insert genomic DNA libraries highly enriched for microsatellite repeat sequences. Proc. Natl. Acad. Sci. USA 89:3419-3423.

Ostrander, E. A., G. F. Sprague, Jr., and J. Rine. 1993. Identification and characterization of dinucleotide repeat $(CA)_n$ markers for genetic mapping in dog. Genomics 16:207-213.

Patterson, D. F., M. E. Haskins, P. F. Jezyk, U. Giger, V. N. Meyers-Wallen, G. Aguirre, J. C. Fyfe, and J. H. Wolfe. 1988. Research on genetic diseases: Reciprocal benefits to animals and man. J. Am. Vet. Med. Assoc. 193:1131-1144.

Patterson, D. F., G. A. Aguirre, J. C. Fyfe, U. Giger, P. L. Green, M. E. Haskins, P. F. Jezyk, and V. N. Meyers-Wallen. 1989. Is this a genetic disease? J. Small Anim. Pract. 30:127-139.

RCP (Registry of Comparative Pathology). 1972-1993. Handbook: Animal Models of Human Disease, fascicles 1-19. Washington, D.C.: Registry of Comparative Pathology. Available from RCP, Armed Forces Institute of Pathology, Washington, DC 20306-6000.

Sarmiento, U. M., and R. F. Storb. 1988a. Characterization of class II alpha genes and DLA-D region allelic associations in the dog. Tissue Antigens 32:224-234.

Sarmiento, U. M., and R. F. Storb. 1988b. Restriction fragment length polymorphism of the major histocompatibility complex of the dog. Immunogenetics 28:117-124.

Sarmiento, U. M., and R. F. Storb. 1989. RFLP analysis of DLA class I genes in the dog. Tissue Antigens 34:158-163.

Scott, J. P., and J. L. Fuller. 1965. Genetics and the Social Behavior of the Dog. Chicago: University of Chicago Press. 468 pp.

Shifrine, M., and F. D. Wilson, eds. 1980. The Canine as a Biomedical Research Model: Immunological, Hematological, and Oncological Aspects. Washington, D.C.: U.S. Department of Energy. 425 pp.

Stone, D. M., P. B. Jacky, and D. J. Prieur. 1991. The Giemsa banding pattern of canine chromosomes, using a cell synchronization technique. Genome 34:407-412.

Swindle, M. M., and R. J. Adams, eds. 1988. Experimental Surgery and Physiology: Induced Animal Models of Human Disease. Baltimore: Williams & Wilkens. 350 pp.

Teichner, M., K. Krumbacher, I. Doxiadis, G. Doxiadis, C. Fournel, D. Rigal, J. C. Monier, and H. Grosse-Wilde. 1990. Systemic lupus erythematosus in dogs: Association to the major histocompatibility complex class I antigen DLA-A7. Clin. Immunol. Immunopathol. 55:255-262.

Vriesendorp, H. M., H. Grosse-Wilde, and M. E. Dorf. 1977. The major histocompatibility system of the dog. Pp. 129-163 in The Major Histocompatibility System in Man and Animals, D. Götze, ed. Berlin: Springer-Verlag.

3

Husbandry

This chapter provides guidelines for the care of laboratory dogs. The first section, on housing, details design and construction considerations for facilities that house dogs, as well as for primary enclosures (here defined as cages and pens). The subsection on facilities contains information on buildings, rooms, and outside areas for containment of dogs, and that on environment and environmental control describes mechanisms for controlling the environment and gives the legislatively mandated ranges for temperature, humidity, and ventilation.

The remaining information in this chapter is supplemented by discussions in other parts of this report. For example, Chapter 4 ("Management of Breeding Colonies") contains sections on food for puppies and gestational or lactating dams and on record-keeping for a breeding colony that amplify the sections on food and identification and records in this chapter. Socialization of puppies is also discussed in Chapter 4. Modified primary enclosures and bedding for dogs with specific disorders are described in Chapter 6 ("Special Considerations").

The 1985 amendment to the Animal Welfare Act required the U.S. Department of Agriculture (USDA) to establish standards for exercise for laboratory dogs, and they were established in 1991. A federal court has now found that the regulations concerning exercise for dogs are inadequate and ordered that new regulations be written. This committee has reviewed the available information relevant to exercise, space, and well-being of dogs, and it has found that, as was the case in 1985, it is inadequate to formulate objective standards.

Although knowledge of canine behavior is leading to a consensus that opportunities for social interaction with people, other dogs, or both are important for promoting canine well-being, no similar consensus is available concerning fitness and exercise. Another issue is the notion that a single standard can provide optimal care for all dogs. It is generally recognized that such factors as breed, physical conformation, age, health status, past experiences, and general behavioral characteristics influence what constitutes adequate space and exercise. For example, a dog undergoing a surgical procedure might require a restricted space to limit its activity. Once the dog has recovered from the surgical procedure, a different space and exercise regimen can be implemented. Likewise, the space and type and duration of exercise required for Alaskan sled dogs in working condition is quite different from that required for Shih Tzu and other brachycephalic breeds. Finally, medical research benefits from the availability of dogs with inherited disorders similar to those of humans, and the presence of these disorders in dogs imposes the same types of restrictions that human patients must endure. Unsupervised exercise is often contraindicated in dogs with heart and metabolic diseases. Similarly, the construction and layout of primary enclosures for dogs with such conditions as muscular dystrophy, bleeding disorders, blindness, or Ehlers-Danlos syndrome must be carefully considered to avoid compromising their health and well-being.

The most important objective for those responsible for housing dogs should be to achieve an overall high level of care, rather than to conform rigidly to specific standards. Animal well-being must be assessed case by case by those qualified to do so. The regular evaluation of animal well-being is an important aspect of any husbandry and animal-care program and serves as a measure of the appropriateness of animal-care procedures. Procedures that are ultimately linked to the well-being of the individual are defined as performance standards. The committee strongly recommends that performance standards, coupled with sound professional judgment, be used to develop space requirements and exercise programs for dogs. This committee is firmly convinced that performance standards are ultimately better for each dog's physical and behavioral well-being than engineering standards, which might lack the flexiblity necessary to meet the needs of all dogs.

HOUSING

Facilities

Housing facilities for dogs must be designed and constructed so that they are structurally sound, protect animals from injury, contain animals securely, and prevent entry of other animals (9 CFR 3.1a). Dog facilities vary in size and complexity, depending on their purpose (e.g., holding or breeding), colony size and type (e.g., specific-pathogen-free or conventional), and breed. The design of breeding facilities should address the following:

- The design should facilitate the conduct of research.
- There should be sufficient space for expansion, both for adding animals and for increasing ancillary operations.
- Breeding facilities should have sufficient space to house dams with litters and the progeny.
- The design should promote effective sanitation and husbandry procedures.
- Operation of the facility should be efficient and cost-effective.
- Construction should be economical.

The physical facilities and equipment should be constructed and operated to fulfill the following criteria:

- Contamination from areas adjacent to, but not part of, the facility should be minimized. The locations of equipment washing and sterilizing, food and bedding storage, quarantine, treatment, receiving and shipping, shipping-crate storage, mechanical services, shops, offices, and laboratories should minimize crossovers from soiled or contaminated to clean areas. Clean material and equipment should not come into contact with soiled and contaminated material and equipment.
- There should be sufficient control of temperature, humidity, ventilation, and lighting to provide the animals with appropriate conditions for their comfort and well-being.
- Behavioral well-being should be considered by allowing for visual contact between dogs, social housing, exercise areas, and other appropriate areas.
- The entry of vermin should be prevented.
- Provisions should be made for lunchrooms, locker rooms, and toilets for animal-care personnel.
- Caging equipment and feeding and watering devices should provide a safe environment, make food and water readily available, minimize the opportunity for transmission of diseases and parasites, and make sanitation and sterilization efficient.
- Auxiliary equipment—such as washing machines, cage racks, rolling equipment (e.g., dollies, tables, and carts), and fixed equipment (e.g., cabinets, sinks, and shelving)—should be designed, fabricated, and used to promote maximal sanitation and operating efficiency.

When a dog facility is designed to be part of a larger facility housing other species of animals or part of a multipurpose building with offices and research laboratories, the physical relationships between areas must be carefully planned (NRC, 1985a). Those establishing operating procedures should use the best available information on physiology; nutrition; genetics; behavior; animal breeding, care, and maintenance; colony management (production and research); and disease control.

Dogs can be housed in indoor facilities, outdoor facilities, or a combination of the two (sheltered housing facilities). If the site is exclusively indoors, the only factors that influence site selection are local zoning regulations, the ability to control odors and noise, the availability of appropriate utilities (e.g., sewerage and water) (9 CFR 3.1d), and the proximity to other businesses. Indoor facilities should be constructed and maintained in compliance with CFR, Title 9, Part 3.2 and the *Guide* (NRC, 1985a), as summarized below.

Indoor Facilities

Walls. Exterior walls should be fire-resistant and impervious to vermin. To facilitate cleaning, interior walls should be smooth, hard, and without pits or cracks, and they should be capable of withstanding the impact of water under high pressure and scrubbing with cleaning agents (e.g., detergents) and sanitizing agents (e.g., disinfectants). They should be protected from damage caused by movable equipment.

Ceilings. Ceilings should be smooth, moistureproof, and free of imperfect junctions. Surface materials should be capable of withstanding scrubbing with detergents and disinfectants. Exposed pipes and fixtures are undesirable.

Floors. Floors can be constructed of a variety of materials that are smooth, moistureproof, nonabsorbent, and skidproof; that are resistant to wear and the adverse effects of detergents, disinfectants, acid, and solvents; and that are able to support heavy equipment without being gouged, cracked, or pitted. They should also be easy to clean.

Drainage. Drainage must be adequate to allow rapid removal of water (9 CFR 3.1f). If floor drains are used, they should be constructed and maintained in accordance with the *Guide* (NRC, 1985a). Rim flush drains should be at least 6 in (15.2 cm) in diameter. Porous trap buckets installed in the drains aid in cleaning and screen out solid waste. Floor drains must contain traps that prevent backflow of sewage and gases (9 CFR 3.1f). If unused floor drains are present, they should be closed with gastight seals that are flush with the floor surface.

Doors. All rooms should have doors. External doors should have adequate latches and locks and should be verminproof when closed. If they are left open during warm weather, adequate screening is essential. All door frames should be sealed to walls and partitions with caulking compound or a similar material.

Ports in animal-room doors allow personnel to observe the dogs without entering the rooms, prevent injury to personnel while they are opening doors, and provide a way to verify that room lights are on at appropriate times. Experience has shown that doors at least 42 in (107 cm) wide and 84 in (213 cm) high allow free passage of cages and equipment. The doors should be equipped with locks and kickplates and should be self-closing.

Outside windows. Outside windows and skylights might not be desirable, because they can contribute to unacceptable variations in temperature and photoperiod. Other problems associated with outside windows and skylights include dust and bacteria buildup on frames, drafts, and increased ventilation costs.

Washrooms and sinks. Washing facilities for personnel (e.g., basins, sinks, or showers) must be provided and must be readily accessible (9 CFR 3.1g).

Sheltered Housing Facilities

A sheltered housing facility, as defined by the Animal Welfare Regulations (AWRs), is a facility that provides shelter, protection from the elements, and protection from temperature extremes at all times (9 CFR 1.1). It can consist of runs or pens in a totally enclosed building or indoor-outdoor runs with the indoor runs in a totally enclosed building. The requirements for the sheltered portion of such facilities are identical with those for indoor facilities, with the additional stipulation that the shelter structure must be large enough to permit each animal to sit, stand, and lie down in a normal manner and to turn around freely (9 CFR 3.3).

The outdoor portion of a sheltered housing facility should be constructed to prevent the introduction of vermin. Outdoor floor areas in contact with animals should be constructed of hard, moisture-resistant material and be properly drained. The use of compacted earth, sand, gravel, or grass is discouraged. The sides of runs can be constructed of chain-link fencing and steel posts or pipe frames or, when necessary to prevent fighting or injury, of solid concrete block coated with sealant. Fencing at the lower ends of runs and pens should be high enough above the surface to permit adequate drainage but not high enough to allow young puppies to escape. Curbs at least 6 in (15.2 cm) high should be constructed between runs to help prevent the spread of microorganisms during washing. Curbs 24-30 in (61.0-76.2 cm) high might be necessary in runs in which the dog population is constantly changing. Higher curbs might be beneficial in whelping-pen runs to reduce the anxiety of nursing bitches. Run doors or gates should have well-made latches that can be easily opened by animal-care personnel but not by the dogs. Special consideration must be given to removing animal wastes and controlling noise.

Outdoor Facilities

The AWRs, with some restrictions, permit facilities to house dogs solely outdoors, provided that each animal has access to a structure (consisting of a roof, four sides, and a floor) that furnishes adequate protection from cold, heat, the direct rays of the sun, and the direct effects of wind, rain, and snow (9 CFR 3.4). In general, this type of housing is discouraged for dogs being used in an experimental protocol, because environmental factors, infectious agents, and vermin are difficult to control. In other instances (e.g., in protocols requiring acclimation or in breeding colonies maintained in temperate climates), outdoor facilities might be adequate.

Environment and Environmental Control

An important part of maintaining the health and well-being of laboratory animals is control of the environment. In nature, animals respond to environmental changes both behaviorally and physiologically in a manner that will maintain homeostasis. In an animal room, a behavioral response might not be possible, and the animal must deal with an altered environment physiologically. Therefore, it is necessary to control the environment to avoid physiologic changes. Besch (1985) has reviewed environmental factors that can effect the biologic responses of laboratory animals.

Temperature and Humidity

Temperature and humidity are important considerations in a dog facility (Besch, 1985). Dogs can tolerate moderate ranges of temperature and weather, provided that they have appropriate amounts of food and water, have access to shelter, and are allowed sufficient time to acclimate to their environment. The *Guide* recommends that room temperature for dogs be maintained within a range of 18-29°C (64.4-84.2°F) and relative humidity within a range of 30-70 percent. The AWRs require that the ambient temperature in indoor facilities not fall below 7.2°C (45°F) or rise above 29.4°C (85°F) for more than 4 consecutive hours when dogs are present (9 CFR 3.2a). Except as approved by the attending veterinarian, ambient temperature must not fall below 10°C (50°F) for dogs not acclimated to lower temperatures, breeds that cannot tolerate lower temperatures, and young, old, sick, or infirm dogs (9 CFR 3.2a).

Dogs recovering from general anesthesia are frequently hypothermic. Every attempt should be made to maintain normal body temperature during surgery and recovery. This can be accomplished by using supplemental sources of heat (e.g., heating pads and heat lamps), by avoiding direct contact with heat-conducting surfaces (e.g., metal), and by maintaining the postoperative recovery cage at 27-29°C (80.6-84.2°F) (NRC, 1985a). Newborn pups have poorly developed thermoregulatory mechanisms and might require supplemental sources of heat. Temperatures of 29.4-32.2°C (85-90°F) have been suggested for the first week of life (Poffenbarger et al., 1990).

Each room should be provided with temperature controls and high- and low-temperature alarms. Graphic recorders are useful for monitoring system performance. Ideally, the temperature controls should allow individual adjustments in dry-bulb temperature of ± 1°C (± 2°F) within the range of 18.3-29.4°C (65-85°F).

Relative humidity should be maintained at 30-70 percent throughout the year (NRC, 1985a). It is important to control sources of humidity, such as cage-cleaning equipment, transient loads from cleaning water (Gorton and Besch, 1974), and thermal and mass loads from animals (Besch, 1991). Low humidity can contribute to respiratory distress; and coughs, pneumonitis, and other problems can follow. High humidity impairs efficient body-cooling (Besch, 1991).

Ventilation

Ventilation serves multiple functions. It supplies oxygen; removes heat generated by animals, lights, and equipment; dilutes gaseous contaminants; and helps to control the effects of infiltration and exfiltration (Clough and Gamble, 1976; Edwards et al., 1983). Gorton et al. (1976) have reported a method for estimating laboratory animal heat loads.

Indoor facilities must be sufficiently ventilated when dogs are present to provide for their comfort and well-being and to minimize odors, ammonia concentrations, drafts, and moisture condensation. Auxiliary ventilation must be provided when the ambient temperature is 29.5°C (85°F) or higher (9 CFR 3.2b). It is commonly thought that 10-15 volumetric changes per hour with outside air must be provided to animal rooms and that air must not be recirculated. As a consequence, animal facilities are generally ventilated with "one-pass" air, although the *Guide* (NRC, 1985a) includes provisions for alternative methods of providing equal or more effective ventilation. Besch (1992) has reviewed alternative methods of ventilation.

Ventilation system design and construction considerations include the following:

- Diffusers and exhaust openings should be located and controlled to prevent drafts.
- Outside openings and exhaust-ventilation grillework should be screened to prevent entry of vermin. Screening should be cleaned regularly.
- Air pressure in clean areas and animal rooms should be greater than that in public and refuse areas. Where pathogenic organisms are present, a negative-pressure system is necessary.
- Ventilating mechanisms should be equipped with suitable alarm systems that will be activated if the temperature moves outside the desired range or if power fails.
- Supplemental exhaust fans or exhaust systems increase drying and reduce humidity when fixed equipment is being washed. If such systems are used, they should be permanently mounted in external windows or wall openings, their frames should be sealed to the building structure, and the systems should be screened.
- Emergency power sources should be available in case of power failure.

Power and Lighting

Electric systems should be safe, furnish appropriate lighting, and provide a sufficient number of outlets. Lighting systems should allow for either manual or timer-controlled changes in illumination levels or photoperiods, and timer performance should be checked regularly. Lighting fixtures, switches, and outlets should be sealed to prevent entry or harboring of vermin. Moistureproof switches and outlets should be installed where water is used in cleaning. Emergency power should be available.

Illumination must be adequate and uniformly diffuse throughout each animal room to allow proper cleaning and housekeeping, to permit inspection of animals, and to maintain the animals' well-being (9 CFR 3.2c). Light levels of 323 lx (30 ft-candles), measured 1.0 m (3.3 ft) above the floor, appear to provide sufficient illumination for routine animal care (Bellhorn, 1980; NRC, 1985a). A regular diurnal lighting cycle must be provided (9 CFR 3.2c).

Noise Control

Barking dogs can be a nuisance both to personnel working in animal facilities and to the adjacent population. Self-generated noise of 80-110 dB (Peterson, 1980; Sierens, 1976) has been measured in dog rooms. The effects of noise on animals are reviewed in the *Guide* (NRC, 1985a).

Noise-control measures should be implemented in both indoor and outdoor environments. Sound transmission can be reduced by using concrete to build walls, covering concrete walls with sound-attenuating material, and eliminating windows (NRC, 1985a). Pekrul (1991) has discussed other means of decreasing noise in animal facilities. Sound-attenuating materials may be bonded to walls or ceilings only if they can be sanitized and will not harbor vermin. Outdoor runs must be designed and constructed to comply with local noise ordinances.

Chemicals and Toxic Substances

Many of the chemicals used in animal facilities for cleaning, sanitizing, pest control, and other purposes can be toxic to housed animals and personnel. In addition, some materials used in construction for coating surfaces can react with certain cleaning and sanitizing agents to produce toxic gases, including chlorine. Where possible, the use of chemicals should be avoided. For example, adequate ventilation is more effective than chemicals in eliminating most animal-room odors, provided that air inlets are not placed near the building exhaust. Newberne and Fox (1978) and Besch (1990) have reviewed chemicals and other toxicants found in animal facilities.

Where chemical agents must be employed, it is essential to be familiar with their potential toxicity and to develop procedures for using and disposing of them properly. Noxious chemicals should not be used to clean animal facilities. Adequate rinsing is essential to prevent the skin irritation or allergic reactions that can be caused by some cleaning and sanitizing agents (e.g., pine oil).

Primary Enclosures

Primary enclosures should facilitate research while maintaining the health and well-being of the dogs. They must confine dogs securely, enable them to remain clean and dry, protect them from injury, and contain sufficient space to allow them to sit, lie, stand, turn around, and walk normally (9 CFR 3.6a). The design should allow inspection of cage or pen occupants without disturbing them and provide easy access to feeding and watering devices for filling, changing, cleaning, and servicing.

Cages or pens should be fabricated of smooth, moisture-impervious, corrosion-resistant materials that can be easily sanitized and sterilized. Floors must be constructed to preclude entrapping toes, dew claws, or collars. Expanded metal or plastic-covered metal mesh is satisfactory for pens or runs, provided that the dogs' feet cannot pass through the openings (9 CFR 3.6a2x). Pen floors must have adequate drainage.

Each cage and pen should have a hinged or sliding door that covers the opening sufficiently to prevent escape of the occupants. Each door should have a latch that holds the door securely closed.

Space Recommendations

The AWRs require that the floor space for each dog equal at least the "mathematical square of the sum of the length of the dog in inches (measured from the tip of its nose to the base of its tail) plus 6 inches [15.24 cm]," expressed in square feet (9 CFR 3.6c1i). In addition, the interior height of each enclosure must be "at least 6 inches [15.24 cm] higher than the head of the tallest dog in the enclosure when it is in a normal standing position" (9 CFR 3.6c1iii). Each bitch with nursing pups must be given additional floor space based on breed and behavioral characteristics and in accordance with generally accepted husbandry practices, as determined by the attending veterinarian (9 CFR 3.6c1ii). The additional space for each nursing pup must be at least 5 percent of the minimum required for the bitch, unless otherwise approved by the attending veterinarian (9 CFR 3.6c1ii). Minimal space recommendations for dogs are also given in the *Guide* (NRC, 1985a, p. 14). These requirements and recommendations are based primarily on professional judgment and convention.

The few scientific studies on this subject have focused on how enclosure size affects movement, activity patterns, and physical fitness. Clark et al. (1991) found no decreases in physical fitness, as measured by heart rate and muscle enzyme (succinate dehydrogenase) activity, when dogs were housed in cages or runs of various sizes that complied with federal standards and guidelines; however, modest decreases in fitness were found when dogs were housed in cages smaller than mandated by the AWRs. It has been shown that, in general, dogs are more active in pens and runs than in cages; however, dogs housed in the largest enclosures are not always the most active (Hetts et al., 1992; Hite et al., 1977; Hughes and Campbell, 1990; Hughes et al., 1989; Neamand et al., 1975). Enclosure size has not been demonstrated to affect the musculoskeletal system (Newton, 1972), cortisol concentrations (Campbell et al., 1988; Clark et al., 1991), or selected measures of immune function (Campbell et al., 1988). Although they provide interesting and relevant information, the studies do not provide sufficient objective, scientific data on which to base space requirements for dogs.

To set standards based on scientific data, one must show a correlation between cage size and behavioral well-being. That poses two problems: it is not clear how to define and measure behavioral well-being, and the determination of well-being depends on human interpretations of the data. Movement and activity patterns are unlikely to be sensitive behavioral measures, because a dog's activity can be increased without improving its well-being (e.g., if there is locomotor stereotypy or increased activity caused by social isolation or competition for space). Moreover, the definition of movement varies between studies, so it is difficult to compare and interpret results. It is generally accepted that a variety of perspectives are needed to assess well-being, including measures of physical health, of neuroendocrine and immunologic responses to stress, of the ability to respond effectively to social and nonsocial environments, and of behavior. Scientific data on dogs are inadequate to support any such assessment relative to enclosure size.

EXERCISE AND ENVIRONMENTAL ENRICHMENT

The requirements for providing opportunities for dogs to exercise are specified in the AWRs (9 CFR 3.8). The following paragraph summarizes the AWRs now in effect. It is incumbent on the reader to keep abreast of changes that might occur as the result of further federal court or USDA actions.

Dogs over 12 weeks old, except bitches with litters, must be given the opportunity for regular exercise if they are kept individually in cages, pens, or runs that are less than 2 times the AWR-required floor space. Dogs housed in groups do not require exercise periods, provided that the total floor space of the cages, pens, or runs equals the sum of the AWR-required spaces for the dogs if housed individually. If a dog is housed without sensory contact with other dogs, it must receive positive physical contact with humans at least once a day. Forced-exercise programs (e.g., swimming or walking on treadmills or carousel devices) are not considered to comply with the AWRs. Each institution is responsible for developing a plan for providing exercise. The plan must be approved by the attending veterinarian and must be made available to USDA on request. Exceptions to the requirement for exercise can be made by the attending veterinarian case by case or, if exercise is inappropriate for a scientific protocol, by the institutional animal care and use committee (IACUC). In the former instance, the exemption from exercise must be reviewed every 30 days, unless it was granted because of a permanent condition (9 CFR 3.8d). In the latter instance, exemptions must be reviewed at appropriate intervals, as determined by the IACUC, but not less often than every 6 months (9 CFR 2.31)

Recent studies have provided some information on exercise and well-being. Clark et al. (1991) and Hetts et al. (1992) found that 30 minutes of forced treadmill exercise five times a week did not affect physical fitness or behavior as measured in the study. Campbell et al. (1988) reported that releasing dogs either singly or as a group into a large area for 35-minute exercise periods three times a week did not affect cage activity patterns or weekly measures of selected hematologic or serum biochemical values. However, dogs were more active during the release periods than in their cages, and dogs released individually had different activity patterns from those of dogs released in groups. Studies on enclosure size and exercise are cited in the section above on space recommendations. Although the studies have provided important and relevant information, sufficient data are still not available to support definitive conclusions about the relationship between exercise and well-being. Future studies should be based on larger samples, use a variety of behavioral measures to evaluate well-being (activity patterns are not likely to be sensitive indicators of well-being), and consider the substantial individual variations in physiologic characteristics that have been reported.

It is well known that dogs are highly social animals, and social isolation and solitary housing are considered to be important stressors of social species (Wolfle, 1990). Solitary housing has been shown to be associated with less activity and with nonsocial repetitive behaviors (Hubrecht et al., 1992). Hetts et al. (1992) have found that socially isolated dogs (i.e., dogs having only auditory contact with other dogs and contact with people only during routine husbandry procedures) display bizarre movement patterns and tend to vocalize more than dogs that have more social contact. Several studies have reported that dogs are more active in the presence of humans (Campbell et al., 1988; Hetts et al., 1992; Hughes and Campbell, 1990; Hughes et al., 1989), especially when human presence is relatively rare (Hubrecht et al., 1992). It has also been shown that dogs housed in pairs sleep more than dogs housed singly (Hetts et al., 1992). Although the relationship between sleep patterns and well-being has not been studied in dogs, there is evidence in other species that normal sleep can be disrupted by a variety of environmental stressors and that return to normal sleep patterns can be a sensitive indicator of an animal's adaptation to environmental changes (Ruckenbusch, 1975).

Evidence of the importance of social interactions for dogs is strong enough to support a recommendation that dogs be socially housed in compatible groups, be given opportunities for social interaction during the exercise period, or both. The AWRs address the compatible grouping of dogs in the same primary enclosure (9 CFR 3.7). Age, sex, experience, and genetic differences in social behavior between individuals and breeds influence how dogs accept social housing and respond to social interaction (Fuller, 1970;

King, 1954; Scott and Fuller, 1965). Social interactions should minimize fearful and aggressive behaviors.

Examples of plans that provide social interactions are leash walking and release of dogs in an enclosed area for specified periods. In the latter, several compatible dogs that are housed in the same room can be released together; however, females in proestrus or estrus should not be released with males. Exercise rooms should be cleaned and sanitized between uses by dogs from separate rooms to minimize disease transmission. Only dogs of similar microbiologic status should be combined in groups (see Chapter 5).

If dogs are to be group-released, the composition of the group should remain as stable as possible (i.e., the members of the group should be the same dogs each time), because how readily a group of dogs accepts new members varies a great deal. Some dogs form closed social groups and attack new members (King, 1954). Changes in group composition often cause instability in the social dominance hierarchy, which in turn can result in intraspecific aggression. It is important to remember that two dogs make a pack, and the behavior of a pack is often very different from that of an individual dog. A thorough understanding of pack structure and social behavior is important for those managing research dogs. Any dog that is being attacked or threatened by the group to the extent that it cannot move about freely should be removed and given an alternative method of exercise. Group-released dogs should be observed frequently during the exercise period to ensure their safety.

Positive social interactions with humans can be achieved by having one or more people in the room during the exercise period. There is evidence that passive contact with a person is more reinforcing to dogs that have been socially isolated than is active contact (Stanley, 1965; Stanley and Elliot, 1962). If a dog displays fearful behavior when handled or petted, the handler should sit passively, avoid eye contact, and allow the dog to approach at will. As fearful behavior decreases, contact can gradually become more active.

Information on other types of environmental enrichment for dogs is scarce. The need for complex or varied environments has not been studied. Dogs have been observed to manipulate and direct attention to loose objects they find in their enclosures (Hetts et al., 1992), and dogs provided with toys spent an average of 24 percent of their time using them (Hubrecht, 1993). The toys reduced the dogs' inactive time and decreased destructive behavior aimed at cage apparatuses (Hubrecht, 1993). The relevance of these behavioral changes to well-being is not yet known. Nonetheless, such devices as balls, chew toys, and ropes might be considered for dogs in restricted environments. It is recommended that an ethologist, comparative psychologist, or animal behaviorist knowledgeable about dog behavior be consulted by those designing exercise and social interaction plans or when other questions arise concerning the behavioral well-being of dogs.

FOOD

Selecting Optimal Rations

Many commercially available dog foods contain all essential nutrients in their required proportions, as outlined in *Nutrient Requirements of Dogs* (NRC, 1985b) and the Association of American Feed Control Officials' *Official Publication 1993* (AAFCO, 1993). These foods are manufactured in dry, semimoist, and canned forms. Dogs should be fed only complete and balanced diets. Specific procedures should be followed to ensure that stored foods do not become deficient in nutrients (NRC, 1985a).

Diet quality can be evaluated by examining the label for a statement of nutritional adequacy, which must be present on all dog-food products sold across state lines. This statement informs the purchaser whether the product has been approved for use as a complete ration for specified life stages (i.e., growth, maintenance, or pregnancy and lactation). Approval is obtained by one of the following means:

• Each of the diet's individual ingredients is analyzed for all essential nutrients; the sum of these nutrients in all ingredients must meet or exceed the nutritional requirements of the animal for specified life stages.
• The product itself is chemically analyzed and shown to meet or exceed the essential-nutrient requirements for specified life stages.
• The product passes a feeding trial as specified by the Association of American Feed Control Officials.

If the product fails to be approved, it must be labeled for use as a dietary supplement only and is not appropriate for use as a dog food. Of the three means of approval, only the feeding trial evaluates the availability of the nutrients in the product. Dog foods approved by that method should be used whenever possible. If such a diet cannot be used, because it would interfere with the experimental design (e.g., nutritional studies with purified diets), the manufacturer of the diet to be used should be consulted about experience with the diet's performance under given conditions.

Many commercially available dog foods, although designed for a specified life stage, are approved and adequate for use during all life stages. Most growth formulations will meet the requirements for gestation, lactation, and maintenance. Similarly, most gestation-lactation products also meet requirements for growth and maintenance. Some foods intended for maintenance will meet the criteria for more than one life stage. However, no food should be used for growth, gestation, and lactation unless its label states that it meets or exceeds nutrient requirements for these life stages.

Special therapeutic diets are available for dogs with specific nutrient requirements caused by the presence of disease (Kirk and Bonagura, 1992; Lewis et al., 1987). Such diets should be fed only under the supervision of a veterinarian.

Feeding

Most commercial rations are formulated to meet all nutrient requirements if a dog eats enough to fulfill its caloric requirements. Estimates of daily caloric requirements can be obtained from several sources, including the manufacturer of the specific food being used. These estimates can be used to initiate feeding programs, but they might need substantial modification because of variations in metabolic rates of individual dogs.

Under most kennel conditions, meal feeding is preferable to free-choice feeding, and individual feeding is preferable to group feeding for the following reasons:

• Restricted feeding has been shown to decrease the incidence of metabolic bone disease in growing dogs that mature at greater than 30 lb (Kealy et al., 1992).

• Restricted feeding has been shown to decrease the incidence of obesity in young beagles and Labrador retrievers (Kendall and Burger, 1980).

• The continual ingestion of small amounts of food observed in free-choice feeding programs stimulates oral bacterial growth and might promote dental disease and gingivitis (Dr. John Saidla, Department of Clinical Sciences, New York State College of Veterinary Medicine, Cornell University, Ithaca, N.Y., unpublished).

• When dogs are fed in groups, dominant dogs might overeat and might prevent subordinate dogs from eating enough to fulfill their daily needs.

• When dogs are fed individually, their food intake can be monitored.

Some kennels have successfully used free-choice feeding to maintain dogs. This practice is most successful when the diet used is a food of relatively low energy density and palatability.

Dogs must be fed at least once a day, except as required for adequate veterinary care (9 CFR 3.9a). Each healthy adult dog should be fed enough to maintain its optimal body weight; this amount will vary with the environment and with the dog's age, sex, breed, temperament, and activity. Within an individual breed, there is often a wide variety of *normal* sizes. It is better to evaluate a dog's size according to how it looks and how it feels than according to body weight alone. With the hands-on approach, a dog's rib cage, spinous processes, and ileal wings should be easily palpable. They should not protrude from under the skin, nor should they be buried under a layer of adipose tissue. Once an adult dog is being maintained at its ideal body size, its weight can be used as a reference for future evaluation of food requirement. However, the loss of muscle mass and gain of adipose tissue, such as are observed in several endocrine disorders, and shifts in fluid balance might make body weight an inaccurate means of assessing nutritional status; therefore, body weight should not completely replace appearance and feel as assessment methods.

Contaminants

Animal-colony managers should be judicious in purchasing, transporting, storing, and handling food to ensure that it does not introduce diseases, potential disease vectors, or parasites. Food must be stored in a manner that prevents spoilage, contamination, and vermin infestation. Open bags must be stored in leakproof containers with tightly fitting lids (9 CFR 3.1e; NRC, 1985a).

Contaminants in food can have dramatic effects on biochemical and physiologic processes. In general, food for dogs should not be manufactured or stored in facilities used for farm foods or any products containing additives, such as rodenticides, insecticides, hormones, antibiotics, fumigants, or other potential toxicants.

WATER

Ordinarily, all dogs should receive fresh, clean, potable water ad libitum. If water is not continuously available, the AWRs require that it be made available at least twice a day for at least 1 hour each time, unless it is restricted by the attending veterinarian (9 CFR 3.10).

Watering devices can be either portable or self-watering. Self-watering devices are convenient and reduce labor, but they require scheduled observations to ensure proper function. Portable watering devices should be easily removable for daily rinsing and periodic sanitizing.

BEDDING AND RESTING APPARATUSES

Bedding can be used in some husbandry situations. For example, if drains are not available, it can be used as an absorbent to help to keep dogs clean and dry. Kinds of bedding typically used for dogs are wood shavings and shredded paper. Bedding must be stored in a manner that protects it from contamination and vermin infestation (9 CFR 3.1e).

Resting apparatuses, especially those made of high-density polyethylene (Britz, 1990), are useful for minimizing loss of body heat from dogs in postoperative recovery, dogs in ill health, and young pups with poorly developed heat-control mechanisms.

SANITATION

The schedule for cleaning and disinfecting dog facilities will vary according to the physical makeup of pens, cages, or runs and other factors. Generally, primary enclosures should be cleaned as needed and sanitized at least once every 2 weeks. Excrement pans and runs should be cleaned daily. If pens and runs composed of materials that cannot be sanitized (e.g., gravel, sand, or pea stone) are used, the contaminated materials should be replaced as often as necessary to prevent odors, diseases, and vermin infestation. Procedures outlined in the AWRs (9 CFR 3.11) should be followed. Dogs must be removed before the floors of primary enclosures are thoroughly cleaned. Primary enclosures containing bitches near parturition, dams with litters, or dogs in quarantine require a cleaning schedule that disturbs them as little as possible.

Equipment and peripheral areas should be cleaned according to the recommendations of the *Guide* (NRC, 1985a). Waste should be removed regularly and frequently, and safe, sanitary procedures should be used to collect and discard it (NRC, 1985a).

IDENTIFICATION AND RECORDS

Identification

Each dog held in a research facility must be marked either with the official USDA tag or tattoo that was on the dog at the time it was acquired or with a tag, tattoo, or collar applied by the facility that individually identifies the dog by number (9 CFR 2.38g1).

Unweaned puppies need not be individually numbered as long as they are maintained in the same primary enclosure as their dam (9 CFR 2.38g3). However, they can be marked for identification with a variety of methods. Colored yarns or spots made with such marking substances as nail polish or paint provide a quick visual reference. Subcutaneous dots can be made by injecting a small amount of tattoo ink beneath the abdominal skin with a tuberculin syringe and 25-gauge needle. Ink dots should be placed in a different location for each pup (e.g., left axilla and right side of abdomen). The location or pattern of the dots and the sex and markings of each pup provide individual identification until permanent tattoos can be applied.

Tattooing of the inner surface of a dog's ear is common. Before the tattoo is applied, the ear should be cleaned thoroughly. Tattoos can be applied with special pliers or an electrovibrator. A tattoo might have to be reapplied after several years. An ancillary method for individually identifying dogs uses a subcutaneously implanted, permanently encoded microchip (transponder) that, when activated by an electronic scanner, broadcasts the encoded number; the scanner transfers the broadcast to a processor that produces either a digital readout or a printed copy. This identification system can be useful during daily examination of dogs being used in studies, but it has not been approved by USDA as the sole source of identification because there is no standard implantation site, no standardized scanner, and no definitive information on whether the microchip migrates from the implantation site. USDA has approved the trial use of the microchips for a few commercial organizations (Richard L. Crawford, Assistant Deputy Administrator for Animal Care, Regulatory Enforcement and Animal Care, APHIS, USDA, Beltsville, Md., personal communication, 1993).

Record-Keeping

Record-Keeping for Scientists and Animal-Care Staff

A life-long, day-to-day log of individual events and experimental procedures experienced by each dog—especially surgery, postsurgical analgesia, and other veterinary interventions—should be carefully maintained. The log will assist animal-care personnel in providing appropriate care, investigators in interpreting research results, and the institution in preparing its annual report to USDA (9 CFR 2.36). Computer programs for maintaining such logs are commercially available (Riley and Blackford, 1991). For small colonies, hand-kept records on each dog might be more appropriate. McKelvie and Shultz (1964) described a record system for long-term studies that is still relevant; it covers clinical examination and includes a coded daily log entry of all events that the animal has experienced.

Records Required by Federal Regulations

Research facilities are obliged to maintain records on procurement, transport, and disposal of all dogs and an inventory of dogs in the facility. When dogs are procured, facilities are required to obtain detailed information on the seller—including name, address, USDA license or registration number or vehicle license number and state—and a description of each dog (9 CFR 2.35b). Likewise, when a dog is transferred to another owner, records must include the name and address of the purchaser, the date and method of transport, and a certificate of health (9 CFR 2.35c). Additional information is available in the section of this chapter entitled "Transportation."

A variety of forms are available to assist institutions in keeping records.

Among them are USDA Interstate and International Certificate of Health Examination for Small Animals (VS Form 18-1), Record of Dogs and Cats on Hand (VS Form 18-5), and Record of Disposition of Dogs and Cats (VS Form 18-6). These forms can be obtained from Regulatory Enforcement and Animal Care, APHIS, USDA, Federal Building, Room 565, 6505 Belcrest Road, Hyattsville, MD 20782 (telephone: 301-436-7833). All records should be maintained for at least 3 years (9 CFR 2.35f).

Records must also be maintained on all offspring born to dogs in the colony (9 CFR 2.35b) and on exceptions to the requirements for exercise (9 CFR 3.8d). Facilities conducting research on any vertebrate animal, including dogs, are obliged to maintain additional records that include the following:

- minutes of meetings of the IACUC;
- semiannual IACUC reports;
- protocols involving animal use;
- scientifically justified deviations from the AWRs; and
- studies involving pain in which analgesics cannot be used.

Some of the information must be reported annually to USDA (9 CFR 2.36); other information, such as approved protocols, must be maintained for 3 years after the study ends (9 CFR 2.35f).

EMERGENCY, WEEKEND, AND HOLIDAY CARE

Dogs should be observed and cared for by qualified personnel every day, including weekends and holidays, as outlined in the *Guide* (NRC, 1985a). Emergency veterinary care should be available after working hours and on weekends and holidays. For dogs undergoing particular experimental procedures and dogs with conditions that might require emergency care, investigators should develop written protocols and provide appropriate additional coverage.

TRANSPORTATION

Transportation over long distances is known to be a stressor for animals. Proper attention to environmental conditions, cage design, and care in transit will minimize the stress. The AWRs specify the requirements for transporting dogs (9 CFR 3.13-3.19). Before a dog is transported, special arrangements must be made between the shipper (consignor), the carrier(s) or intermediate handlers, and the recipient (consignee). The shipper must certify that the dog was offered food and water during the 4 hours before delivery to the carrier and must prepare a written certification, which must be securely attached to the cage and must contain the shipper's name and address, the animal identification number, the time and date when the dog was last offered food and water, specific instructions for feeding and watering the dog for a 24-hour period, and the signature of the shipper with the date and time when the certification was signed.

Primary Enclosures

Carriers must not accept dogs for shipment if their primary enclosures do not meet the requirements of the AWRs (9 CFR 3.14). The primary enclosure must be large enough to allow a dog to turn around while standing, to stand and sit erect, and to lie in a natural position. Primary enclosures must be structurally sound, free of internal protrusions that could cause injury, constructed of nontoxic materials, and able to withstand the normal rigors of transportation. The container must secure the animal and all parts of its body inside the enclosure. Devices, such as handles, must be attached to the outside to allow the container to be lifted without tilting. The container must have a leakproof, solid floor or have a raised floor and a leakproof collection tray. If animals are housed directly on the floor, absorbent bedding material must be provided. Primary enclosures must be cleaned and any litter replaced if dogs are in transit for more than 24 hours. Primary enclosures should be well ventilated to minimize the potential for a thermal gradient during shipment. Additional specifications for transport cages are in the AWRs (9 CFR 3.14) and the *IATA Live Animal Regulations* (IATA, 1993 et seq.).

Puppies 4 months old or younger must not be transported in the same primary enclosure with adult dogs other than their dams. For puppies shipped during sensitive periods of behavioral development (i.e., 8-14 weeks of age; see Scott and Fuller, 1965), shipping stress should be minimized. Dogs likely to display aggressive behavior must be shipped individually, and females in heat must not be transported in the same primary enclosures as males. No more than two live puppies 8 weeks to 6 months old, of comparable size, and weighing 9 kg (20 lb) or less each may be transported by air in the same primary enclosure. Older dogs and puppies weighing more than 9 kg (20 lb) should be individually housed. Weaned littermates that are less than 8 weeks old and are accompanied by their dam may be transported in the same enclosure to research facilities, either by air or surface transport. During transport by surface vehicle, no more than four dogs 8 weeks old or older and of comparable size may be transported in the same primary enclosure.

When viral-antibody-free (unvaccinated) dogs are transported between facilities, precautions must be taken to avoid contact with infectious agents. Some commercial suppliers have developed filtered shipping containers to transport those dogs. IATA rules require that special measures be taken to ensure that ventilation rates are maintained within the container, that the container be appropriately labeled, that sufficient water be provided for the entire journey, and that food, if required, be provided at the point of origin (IATA, 1993).

Environmental Conditions

At all times, containers holding dogs should be placed in climate-controlled areas that provide protection from the elements (9 CFR 3.13, 3.15, 3.18-3.19). Trucks and planes must be ventilated and provide air that has adequate oxygen and is free of harmful gases and particulate contaminants. Airlines should always place dogs in pressurized compartments. Dogs may be shipped if temperatures will fall below 7.2°C (45°F) during any portion of their journey only if a veterinarian certifies in writing that they have been acclimated to lower temperatures and states the lowest temperature to which they have been acclimated. During transit, dogs must not be exposed to ambient temperatures exceeding 29.4°C (85°F) for a period of more than 4 hours.

Food and Water

All dogs must be offered food and water within 4 hours of delivery to the carrier (9 CFR 3.13c). Carriers must offer water to each dog at 12-hour intervals beginning 12 hours after the shipper last offered water. Adult dogs must be fed at least once every 24 hours, and puppies less than 16 weeks old must be fed every 12 hours throughout the trip. Feeding and watering utensils must be firmly secured to the inside of the container and placed so that they can be filled from outside the container. Written instructions for feeding and watering in transit must be attached to the primary enclosure in such a way that they are easily seen and read (9 CFR 3.16).

Other Requirements

There are special requirements for animal holding areas of terminal facilities, including rules for sanitization, pest control, ventilation, temperature control, and shelter from direct sunlight, rain, snow, and extreme heat (9 CFR 3.18).

Each dog must be accompanied by a health certificate, issued by a licensed veterinarian not more than 10 days before shipping, that states that the dog is free of any infectious disease or physical abnormality that would endanger it or other animals or pose a threat to public health. An exemption can be made by the secretary of USDA for individual animals shipped to research facilities if the facilities require animals that are not elegible for certification (9 CFR 2.78b). Instructions for the administration of drugs or provision of other special care must be firmly attached to the outside of the container (9 CFR 3.14h). A pregnant bitch should be accompanied by a certificate, signed by a veterinarian, that states that there is no risk of birth during transit (IATA, 1993).

Carriers and intermediate handlers must not accept dogs more than 4 hours before the scheduled departure (6 hours by special arrangement). An attempt must be made to notify the recipient on arrival at the destination and at least once every 6 hours thereafter (9 CFR 3.13f). During shipment by surface transportation, the operator of the conveyance or someone accompanying the operator must observe the dogs at least once every 4 hours to ascertain that they have sufficient air for normal breathing and are not in distress and that the rules for ambient temperature and all other AWR requirements are met. The same rules apply in air carriers if the animal cargo area is accessible during flight. If it is not accessible, the carrier must observe the dogs at loading and unloading. Dogs in physical distress must receive veterinary care as soon as possible (9 CFR 3.17).

REFERENCES

AAFCO (Association of American Feed Control Officials), Canine Nutrition Expert Subcommittee, Pet Food Committee. 1993. AAFCO nutrient profiles for dog foods. Pp. 92-99 in Official Publication 1993. Atlanta: Association of American Feed Control Officials. Available from Charles P. Frank; AAFCO Treasurer; c/o Georgia Department of Agriculture; Plant Food, Feed, and Grain Division; Capitol Square, Atlanta, GA 30334.

Bellhorn, R. W. 1980. Lighting in the animal environment. Lab. Anim. Sci. 30(2): 440-450.

Besch, E. L. 1985. Definition of laboratory animal environmental conditions. Pp. 297-315 in Animal Stress, G. P. Moberg, ed. Bethesda, Md.: American Physiological Society.

Besch, E. L. 1990. Environmental variables and animal needs. Pp. 113-131 in The Experimental Animal in Biomedical Research. Vol. I: A Survey of Scientific and Ethical Issues for Investigators, B. E. Rollin and M. L. Kesel, eds. Boca Raton, Fla.: CRC Press.

Besch, E. L. 1991. Temperature and humidity control. Pp. 154-166 in Handbook of Facilities Planning. Vol. 2: Laboratory Animal Facilities, T. Ruys, ed. New York: Von Nostrand Reinhold.

Besch, E. L. 1992. Animal facility ventilation air quality and quantity. ASHRAE Trans. 98(pt. 2):239-246.

Britz, W. E., Jr. 1990. Caging systems for dogs under the new standards of the Animal Welfare Act. Pp. 48-50 in Canine Research Environment, J. A. Mench and L. Krulisch, eds. Bethesda, Md.: Scientists Center for Animal Welfare. Available from SCAW, 4805 St. Elmo Avenue, Bethesda, MD 20814.

Campbell, S. A., H. C. Hughes, H. E. Griffin, M. S. Landi, and F. M. Mallon. 1988. Some effects of limited exercise on purpose-bred beagles. Am. J. Vet. Res. 49:1,298-1,301.

Clark, J. D., J. P. Calpin, and R. B. Armstrong. 1991. Influence of type of enclosure on exercise fitness of dogs. Am. J. Vet. Res. 52:1,024-1,028.

Clough, G., and M. R. Gamble. 1976. Laboratory Animal Houses. A Guide to the Design and

Planning of Animal Facilities. LAC Manual Series No. 4. Carshalton, Surrey, U.K.: Medical Research Council Laboratory Animals Centre. 44 pp.

Edwards, R. G., M. F. Beeson, and J. M. Dewdney. 1983. Laboratory animal allergy: The measurement of airborne urinary allergens and the effect of different environmental conditions. Lab. Anim. (London) 17:235-239.

Fuller, J. L. 1970. Genetic influences on socialization. Pp. 7-18 in Early Experiences and the Process of Socialization, R. A. Hoppe, G. A. Milton, and E. C. Simmel, eds. New York: Academic Press.

Gorton, R. L., and E. L. Besch. 1974. Air temperature and humidity response to cleaning water loads in laboratory animal storage facilities. ASHRAE Trans. 80(pt. 1):37-52.

Gorton, R. L., J. E. Woods, and E. L. Besch. 1976. System load characteristics and estimation of annual heat loads for laboratory animal facilities. ASHRAE Trans. 82(pt. 1):107-112.

Hetts, S., J. D. Clark, J. P. Calpin, C. E. Arnold, and J. M. Mateo. 1992. Influence of housing conditions on beagle behaviour. Appl. Anim. Behav. Sci. 34:137-155.

Hite, M., H. M. Hanson, N. R. Bohider, P. A. Conti, and P. A. Mattis. 1977. Effect of cage size on patterns of activity and health of beagle dogs. Lab. Anim. Sci. 27:60-64.

Hubrecht, R. C. 1993. A comparison of social and environmental enrichment methods for laboratory housed dogs. Appl. Anim. Behav. Sci. 37:345-361.

Hubrecht, R. C., J. A. Serpell, and T. B. Poole. 1992. Correlates of pen size and housing conditions on the behaviour of kennelled dogs. Appl. Anim. Behav. Sci. 34:365-383.

Hughes, H. C., and S. Campbell. 1990. Effects of primary enclosure size and human contact. Pp. 66-73 in Canine Research Environment, J. A. Mench and L. Krulisch, eds. Bethesda, Md.: Scientists Center for Animal Welfare. Available from SCAW, 4805 St. Elmo Avenue, Bethesda, MD 20814.

Hughes, H. C., S. Campbell, and C. Kenney. 1989. The effects of cage size and pair housing on exercise of beagle dogs. Lab. Anim. Sci. 39:302-305.

IATA (International Air Transport Association). 1993. IATA Live Animal Regulations, 20th ed. Montreal, Quebec: International Air Transport Association. Available from IATA, Publications Department, 2000 Peel Street, Montreal, Quebec, Canada H3A 2R4.

Kealy, R. D., S. E. Olsson, K. L. Monti, D. F. Lawler, D. N. Biery, R. W. Helms, G. Lust, and G. K. Smith. 1992. Effects of limited food consumption on the incidence of hip dysplasia in growing dogs. J. Am. Vet. Med. Assoc. 201:857-863.

Kendall, P. T., and I. H. Burger. 1980. The effect of controlled and appetite feeding on growth and development in dogs. Pp. 60-63 in Proceedings of the Kal Kan Symposium for the Treatment of Dog and Cat Diseases (Sept. 29-30, 1979), R. L. Wyatt, ed. Vernon, Calif.: Kal Kan Foods, Inc. Available from Kal Kan Foods, Inc., 3250 E 44th Street, Vernon, CA 90058-0853.

King, J. A. 1954. Closed social groups among domestic dogs. Proc. Am. Philos. Soc. 98:327-336.

Kirk, R. W., and J. D. Bonagura, eds. 1992. Current Veterinary Therapy. XI. Small Animal Practice. Philadelphia: W. B. Saunders. 1,346 pp.

Lewis, L. D., M. L. Morris, Jr., and M. S. Hand. 1987. Small Animal Clinical Nutrition III. Topeka, Kans.: Mark Morris Associates. Available from Mark Morris Associates, 5500 SW 7th Street, Topeka, KS 66606.

McKelvie, D. H., and F. T. Shultz. 1964. Methods of observing and recording data in long-term studies on beagles. Lab. Anim. Care 14:118-124.

Neamand, J., W. T. Sweeney, A. A. Creamer, and P. A. Conti. 1975. Cage activity in the laboratory beagle: A preliminary study to evaluate a method of comparing cage size to physical activity. Lab. Anim. Sci. 25:180-183.

Newberne, P. M., and J. G. Fox. 1978. Chemicals and toxins in the animal facility. Pp. 118-141 in Laboratory Animal Housing. Proceedings of a symposium organized by the Institute of Laboratory Animal Resources Committee on Laboratory Animal Housing. Washington, D.C.: National Academy of Sciences.

Newton, W. M. 1972. An evaluation of the effects of various degrees of long-term confinement on adult beagle dogs. Lab. Anim. Sci. 22:860-864.

NRC (National Research Council), Institute of Laboratory Animal Resources, Committee on Care and Use of Laboratory Animals. 1985a. Guide for the Care and Use of Laboratory Animals. NIH Pub. No. 86-23. Washington, D.C.: U.S. Department of Health and Human Services. 83 pp.

NRC (National Research Council), Board on Agriculture, Subcommittee on Dog Nutrition, Committee on Animal Nutrition. 1985b. Nutrient Requirements of Dogs, revised ed. Washington, D.C.: National Academy Press. 79 pp.

Pekrul, D. 1991. Noise control. Pp. 166-173 in Handbook of Facilities Planning. Vol. 2: Laboratory Animal Facilities, T. Ruys, ed. New York: Von Nostrand Reinhold.

Peterson, E. A. 1980. Noise and laboratory animals. Lab. Anim. Sci. 30:422-439.

Poffenbarger, E. M., M. L. Chandler, S. L. Ralston, and P. N. Olson. 1990. Canine neonatology. Part 1. Physiologic differences between puppies and adults. Compend. Cont. Educ. Pract. Vet. 12:1601-1609.

Riley, R. D., and R. K. Blackford. 1991. ALACARTE—An animal in-life tracking system. AALAS Bull. 30(3):20-23. Available from the American Association for Laboratory Animal Science, 70 Timber Creek Drive, Suite 5, Cordova, TN 38018.

Ruckenbusch, Y. 1975. The hypnogram as an index of adaptation of farm animals to changes in their environment. Appl. Anim. Ethol. 2:3-18.

Scott, J. P., and J. L. Fuller. 1965. Genetics and the Social Behavior of the Dog. Chicago: University of Chicago Press. 468 pp.

Sierens, S. E. 1976. The Design, Construction, and Calibration of an Acoustical Reverberation Chamber for Measuring the Sound Power Levels of Laboratory Animals (thesis for M.S. degree). Gainesville: University of Florida. 127 pp. Available from Health Science Center Library, University of Florida, Box 100206, Gainesville, FL 32610-0206.

Stanley, W. C. 1965. The passive person as a reinforcer in isolated beagle puppies. Psychon. Sci. 2:21-22.

Stanley, W. C., and O. Elliot. 1962. Differential human handling as reinforcing events and as treatment influencing later social behavior in basenji puppies. Psychol. Rep. 10:775-788.

Wolfle, T. L. 1990. Policy, program and people: The three P's to well-being. Pp. 41-47 in Canine Research Environment, J. A. Mench and L. Krulisch, eds. Bethesda, Md.: Scientists Center for Animal Welfare. Available from SCAW, 4805 St. Elmo Avenue, Bethesda, MD 20814.

4

Management of Breeding Colonies

REPRODUCTION

To maintain the breeding efficiency of a colony or to breed an important individual dog successfully, staff must understand the unique reproductive characteristics of dogs. The biology of canine reproduction has been extensively reviewed (Burke, 1986; Christiansen, 1984; Concannon, 1991; Concannon and Lein, 1989; Concannon et al., 1989). Information on heritability of physical and other characteristics of dogs, Mendelian genetics of breeding, the incidence and characteristics of diseases that have a genetic basis, and methods for demonstrating heritability is also available (Patterson, 1975; Patterson et al., 1989; Shultz, 1970; Willis, 1989).

Reproductive Cycle of the Bitch

Most bitches can become pregnant once or twice a year. Each ovarian cycle consists of the following phases:

• A follicular phase, or proestrus, during which there is progressive vulval swelling and a serosanguineous (bloody) vaginal discharge. During this period, which can last from 3 days to 3 weeks, the bitch's blood has high concentrations of estrogen. The male will show interest, but he either does not or is not allowed to mount.

• A periovulatory period, or estrus, during which estrogen declines and progesterone increases as the ovarian corpora lutea form. This period is also the early luteal phase of the cycle. During estrus, which can last from 3 days to 3 weeks, the bitch assumes a characteristic posture in the presence of a male in which the rump is raised and there is a curvature of the back (lordosis) and the tail is held to one side (flagging). The male is allowed to mount, and copulation occurs.

• A midluteal and late luteal phase or metestrus (either pregnant or nonpregnant metestrus), which lasts about 2 months and during which serum progesterone remains elevated above 1 ng/ml.

• A period of weak ovarian activity, or anestrus, lasting 2-10 months, in which progesterone concentration is low, and there is no evidence of estrogen stimulation of the uterus or vulva.

In constant photoperiods of 12 hours of light and 12 hours of darkness or 14 hours of light and 10 hours of darkness, estrous periods should occur with equal incidence throughout the year. Possible effects of constant light have not been studied. With natural circannual changes in photoperiod, bitches come into estrus more frequently in winter and spring months than in summer and autumn months. In most breeds, the interval between estrous periods averages 7-8 months. After the age of 8 years, however, the interval between cycles begins to lengthen, reaching 12 months or longer by the age of 12 years (Andersen and Simpson, 1973).

Successful breeding requires that observation of reproductive conditions be given high priority, and staff must be able to recognize the start of proestrus. A swollen vulva might not be obvious on a dark or long-haired dog, and bitches often lick away the bloody discharge; therefore, the vulva of each breeding bitch must be examined closely two or three times a week, beginning 4 months after estrus.

Vaginal cytology can be useful for estimating the best time for breeding (Concannon and DiGregorio, 1986; Holst, 1986; Olson et al., 1984) and predicting the time of whelping, which will be 55-60 days after a change in the smear indicates late estrus. Vaginal smears in anestrus are nondescript, with a few leukocytes and small epithelial cells. In early proestrus, smears include a high proportion of rounded epithelial cells, erythrocytes, and sometimes a few leukocytes. During midproestrus, there is an increasing percentage of cornified (flakelike) epithelial cells but no leukocytes. All or nearly all epithelial cells in the smear are cornified from 2-8 days before ovulation until 4-9 days after ovulation, when these cells predictably and abruptly decline. In early metestrus, cornified cells are replaced by rounded, smaller superficial cells, and there is usually an influx of leukocytes. The metestrus smear slowly regresses to the nondescript anestrus smear. Smears should be taken from the anterior vagina. They should be obtained and prepared carefully with saline-moistened swabs and should not be contaminated with vulval material. In the case of bitches that have had reproductive problems, when a successful breeding is important, or both, more accurate predictions can be made by monitoring the progesterone concentration in the serum or plasma with an enzyme-linked immunosorbent assay (ELISA) kit (Bouchard et al., 1991a; Hegstad and Johnston, 1992; Johnston and Romagnoli, 1991). In this test, ovulation occurs a mean of 1-2 days after the initial rise in progesterone, peak fertility a mean of 0-4 days after the initial rise, loss of fertility 6-11 days after, implantation 18-20 days after, and parturition 63-65 days after (Concannon, 1991).

Mating

Theoretically, it is sufficient to maintain one male for every 10-20 females; however, in practice this ratio might not be adequate, for several reasons. First, a bitch in proestrus produces pheromones that will start proestrus in other bitches in the colony, making it likely that several bitches will be in estrus simultaneously. Because mating an individual male more often than once each day can reduce its sperm output after 1 week (Amann, 1986), a greater ratio of males to females might be required to maintain breeding efficiency. Second, except under special circumstances, such as reproducing a disease model, breeding programs should conscientiously avoid inbreeding, and it has been estimated that a ratio greater than two males for each 10 females is needed to prevent an increase in the coefficient of in-breeding (Shultz, 1970).

Natural Mating

Mating can be done naturally or by artificial insemination with fresh or frozen and thawed semen. Provided that the male is healthy, it is not necessary to take special precautions or to use medications to treat the genitalia because the vagina is not a sterile environment. However, it is important to ascertain that neither the dog nor the bitch has canine brucellosis, a disease that seriously affects reproduction and is a zoonosis (see "Control of Infectious Diseases" in Chapter 5). The bitch is usually taken to the stud dog's pen or cage, because a dog will often ignore the bitch or spend an inordinate amount of time scent-marking if he is moved to new surroundings. The bitch should be mated on 2 or 3 days over a 3- to 5-day period. Unless the staff is experienced in distinguishing early proestrus from estrus, the bitch should be presented to the male for 10-15 minutes every day or every other day from the time she is found to be in proestrus until she is mated. Breeding pairs should not be left unattended, because some bitches are highly selective in choosing mates and it is not uncommon for a bitch to attack a dog that is not of her choosing. In addition, the dog might need assistance until he attains a copulatory lock. Mating should be recorded only on the basis of observations of copulatory locks that last several minutes or more. If a bitch refuses a particular dog, even when signs of estrus (lordosis and flagging) are present, placing her with a different dog might solve the problem. If it is important to the breeding program that a bitch be bred to a dog that she is refusing, a caretaker should restrain her in a manner that will prevent her biting either the caretaker or the stud dog during breeding, or artificial insemination (AI) should be used.

To ensure accuracy of parentage, the same stud must be used for every breeding within a single estrus to avoid multiple-sire litters. Bitches allow dogs to mate from several days before ovulation until several days after ovulation. Because the events of pregnancy are related to the time of ovulation—not necessarily to the time of mating—parturition can occur 56-68 days after a single mating and up to 70 days after the first of multiple matings. Sperm can survive 6 days or more in the bitch, and ovulated eggs can remain fertile for 3-7 days. Parturition should occur 62, 63, or 64 days after ovulation in nearly every bitch (Concannon et al., 1983). Bitches that whelp 56-60 days after the first mating often have small litters, probably because they were bred at the end of the fertile period (P. Concannon, New York State College of Veterinary Medicine, Cornell University, Ithaca, N. Y., unpublished).

Artificial Insemination

AI can be helpful when males cannot be moved easily within or between facilities, when breeding females with weak or selective estrus behavior, when using males that cannot provide natural service, and for preserving valuable animal models. Semen collection, handling of semen, and insemination are described in detail elsewhere (Christiansen, 1984; Concannon and Battista, 1989).

Insemination with fresh semen. Semen can be collected in a clean paper cup or in a latex cone (artificial vagina) attached to a 15-ml conical polypropylene centrifuge tube. An advantage of the former method is that debris from the penis is less likely to become mixed in the ejaculate. Ejaculate should be maintained at room or skin temperature and should be checked microscopically for sperm viability and malformations. Any variation from the expected chalky white color or 1- to 5-cc volume should be recorded. The full ejaculate should be deposited into the anterior vagina with a clean plastic pipet attached to a syringe with nonrubber (e.g., polypropylene) tubing. The hindquarters of the bitch should be raised for 10 minutes while the vagina is manipulated digitally by an attendant wearing a clean glove. The bitch should not be allowed to sit for 20 minutes, and pressure on her abdomen should be avoided. AI should be performed every other day until two or three inseminations have been accomplished. The precise timing for performing AI can be predicted by checking for softening of the vulva, which often occurs around the time of ovulation; by demonstrating the appropriate vaginal cytologic characteristics of advanced estrus; or by measuring the initial rise in serum or plasma progesterone. Ideally, two inseminations should occur before vaginal smears show reduced cornification.

Insemination with fresh chilled semen. Fresh semen can be diluted or extended in one of several laboratory buffers or commercial extenders and shipped refrigerated by overnight express for use in insemination in another location (Concannon and Battista, 1989). At 4°C (39.2°F), sperm motility remains nearly normal for 3-4 days if the semen is diluted in an appropriate diluent and for 1 day if undiluted (see Morton and Bruce, 1989).

Insemination with frozen semen. Frozen semen should be thawed and handled according to the instructions provided by the laboratory that processed it, because each freezing technique has stringent requirements for rate of thawing, dilution, and site of deposition. Although sperm live for several days in fresh semen, they normally die within a few hours after thawing; therefore, precise timing of insemination is important for successful impregnation. The best time to inseminate is usually shortly after oocyte maturation, which occurs 5-6 days after the initial rise in progesterone, around the time of a surge in leutinizing hormone. In most bitches, the inseminations should also take place 2-4 days before the decrease in vaginal cornification. Reported success rates for vaginal insemination range from 0 to 70 percent (Concannon and Battista, 1989); success probably depends heavily on the freezing method and the number of viable sperm inseminated. Success rates of 50-90 percent have been reported for uterine insemination, which is accomplished surgically or with special instrumentation to deposit sperm through the cervix (Concannon and Battista, 1989).

Pregnancy and Parturition

Pregnancy can be determined at 25 days after ovulation by ultrasonography, at 20-35 days after ovulation with palpation, and at 45 days after ovulation with radiography (Johnson, 1986; Yeager and Concannon, 1990). There are no well-documented biochemical or immunologic canine pregnancy tests available. Concannon (1991) has reviewed changes in body weight during pregnancy and pregnancy-specific changes in hematocrit, serum chemistry, and metabolism.

Whelping facilities should provide seclusion from excessive noise and other disturbances. The whelping box should be large enough to accommodate the bitch and pups and have sides high enough to prevent neonates from wandering out of the box. The bottom of a large, fiberglass shipping crate works well for beagle-size dogs. The whelping box should be provided about a week before expected parturition.

Johnson (1986) has reviewed the management of the pregnant bitch. A nonpurulent green discharge, anorexia, and restlessness are normal just before parturition. Birth of a litter can be either rapid or protracted over much of a day. Intervals between pups normally range from 20 minutes to 3 hours. Intervals greater than 3 hours can indicate a problem with fetal position or uterine function and warrant veterinary attention. Persistent, unproductive labor of more than 1 hour also requires veterinary attention (Johnston and Romagnoli, 1991; Jones and Joshua, 1988).

NEONATAL CARE

Newborn pups, like all neonatal mammals, have poorly developed temperature-control mechanisms; therefore, it is necessary to keep the temperature in the whelping box higher than room temperature. Temperatures of 29.4-32.2°C (85-90°F) have been suggested for the first 7 days of life, 26.7°C (80°F) for days 8-28, 21.1-23.9°C (70-75°F) for days 29-35, and 23.9°C (70°F) thereafter (Poffenbarger et al., 1990). That can be done by raising the temperature of the room and placing insulation between the whelping box and the cage or floor or by using heating devices, such as heat lamps or built-in heating elements. However, caution is necessary in using such heating devices; because pups younger than 7 days old have very slow withdrawal reflexes (Breazile, 1978), they can be overheated or severely burned by these devices. Circulating-water heating pads or commercial pig warmers are useful, because they maintain heat at a safe level.

Whelping boxes should be examined two or more times a day for evidence of maternal neglect or cannibalism and for problems with the pups. A normal pup is plump and round, its head is mobile, and it exhibits a rooting reflex. Breathing is regular and unlabored, and the coat is shiny and free of debris. Abdominal enlargement after nursing is normal, but abdominal enlargement accompanied by restlessness, weakness, and either excessive vocalization or complete silence can indicate illness or aerophagia. Failure to gain weight is often the first sign of illness in a newborn animal (Greco and Watters, 1990). Andersen (1970) reported expected weight gains for beagle pups.

Dead pups should be removed from the box. Andersen (1970) and Lawler (1989) have reviewed causes of neonatal deaths and have reported an average rate of death of about 20 percent. Necropsy examination is suggested for all pups that die or are euthanatized with severe illness. Such examinations are necessary to distinguish between congenital defects, which affect only the pups in which they occur; infectious diseases, whose spread might be prevented; and problems with the dam (e.g., insufficient milk) or the environment (e.g., room temperature too low), which can be corrected.

REPRODUCTIVE PROBLEMS

False Estrus and Anestrus

Recurrent frequent false estrus (estrus without ovulation) has been reported (Shille et al., 1984). In false estrus, estrus appears normal, and bitches will mate but fail to conceive. False estrus can be confirmed by demonstrating with a progesterone ELISA kit that the serum or plasma progesterone concentration has not risen above 1 ng/ml, as would be expected for 50 days or more after ovulation if estrus were normal. Bitches that often have false estrus or have false estrus followed in a few weeks by normal estrus cause problems in maintaining breeding colonies. Except in special circumstances, such as reproducing a disease model, it is preferable to cull these animals. Culling based on small litter size, problems with whelping or maternal behavior, chronic infertility, or persistent anestrus is also appropriate. Methods for assessment and treatment for potential causes of infertility in females have been extensively reviewed (Feldman and Nelson, 1987; Johnston and Romagnoli, 1991; Shille, 1986). Persistent anestrus can be distinguished from unobserved cycles only through extremely careful examinations for signs of proestrus or progesterone assays every 6 weeks. Estradiol assays are not particularly informative, and assays of canine gonadotropin to diagnose primary gonadal failure are not readily available. Attempts to induce estrus in anestrus bitches have had variable success (Bouchard et al., 1991b; Concannon, 1992; Concannon et al., 1989).

Delayed Parturition

Whelping should not be considered overdue until 67 days after the last mating or possibly 70 or more days after the first of several matings. Cesarean section should not be contemplated earlier unless there are obvious signs of distress in the bitch. Johnson (1986) and Jones and Joshua (1988) have reviewed veterinary management of dystocia.

Pseudopregnancy

Bitches that are not bred or that are bred but fail to become pregnant frequently exhibit pseudopregnancy because of the progesterone secretion that always follows ovulation. Signs of pseudopregnancy include extensive mammary development, lactation, and maternal behavior. Pseudopregnancy is rare in beagles but more common in other breeds. It is self-limiting and usually does not require intervention (Feldman and Nelson, 1987).

SPECIAL NUTRITIONAL REQUIREMENTS

Bitches

During pregnancy and lactation, bitches should be fed a diet approved by the Association of American Feed Control Officials for all life stages or a diet specially formulated for gestation and lactation (see "Selecting Optimal Rations" in Chapter 3). When a quality diet is fed, supplementation with vitamins and minerals is neither necessary nor desirable.

During the first two-thirds of pregnancy, the amount fed should be the same as that fed before pregnancy. During the last trimester, food intake should be gradually increased so that at parturition it is 150 percent of the daily maintenance requirement. Bitches should not be permitted to become obese during gestation, because this condition can increase the risk of dystocia and postparturient metabolic disorders (Johnston, 1986). Bitches that are underfed during gestation tend to have a higher incidence of stillbirths than bitches that are fed appropriate amounts, and their pups often weigh less at birth (Holme, 1982).

Lactation represents the greatest nutrient challenge that bitches experience during their lifetimes. For the first 3 weeks after parturition, nutrient requirements increase rapidly, leveling off at 200-250 percent of daily maintenance requirements, or even more, depending on the number of nursing pups (NRC, 1985). The nutritional demands of lactation are met best through free access to both food and water. At the time of weaning, food is generally withheld for 24 hours to decrease milk production. Food intake for the first day after weaning should be one-fourth of the amount required for maintenance and then gradually increased to the maintenance requirement by day 4. Ideally, lactating bitches should be within 15 percent of their prebreeding body weight at the time of weaning (AAFCO, 1993).

Pups

Pups should be maintained exclusively on their dams' milk until they are 3 weeks old. They can then begin to eat small amounts of a moistened gestation-lactation diet or a growth diet. Most pups can be weaned completely onto this type of diet by the age of 6-8 weeks. For the development of normal social behavior, it is desirable that they not be completely weaned before they are 6 weeks old. Pups that cannot be nursed by their dam or a foster dam before they are 5 weeks old should be fed one of the commercially available, complete milk replacers. Pups can be fed with bottles and nipples or stomach tubes. Bottles and nipples should be thoroughly cleaned after each use. If a stomach tube is used, its proper placement can be ensured by inserting it to a distance equal to the premeasured distance from the mouth to the last rib. A small amount of sterile saline solution should be introduced through the tube before milk replacer is injected. After each meal, orphaned pups should be massaged in the anal-genital region with a warm, wet cotton ball to stimulate urination and defecation. Most orphans can be completely weaned onto solid food by 5 weeks of age.

Young pups most readily eat canned or moistened dry food; older pups can be fed dry, semi-moist, or canned food. Pups can be fed on either a free-choice or meal-feeding program. If a meal-feeding program is used, they should be fed at least four times a day until they are 3 months old, three times a day until they reach two-thirds of their adult weight, and two times a day thereafter. After the age of 3 months, free-choice programs can lead to obesity in small breeds and faster than optimal growth in large breeds. Excessively rapid growth in breeds whose weight at maturity is more than 30 lb has been associated with an increase in the incidence of several metabolic bone diseases (Hedhammer, 1981; Hedhammer et al., 1974; Kealy et al., 1992). Pups should be fed so that they grow at near optimal rates; growth-curve data are often available from pet-food manufacturers. When an appropriate growth ration is fed, no supplementation is necessary. If a product is not capable of supporting an optimal growth rate, it is generally safer, less expensive, and more convenient to switch to a better-quality growth diet. As a general rule, pups gain approximately 1-2 g/day per pound of anticipated adult body weight (Lewis et al., 1987). An inappropriate growth rate usually reflects a problem with the ration being fed or with the pups' access to it.

VACCINATION AND DEWORMING

Annual vaccinations and deworming of brood bitches should be scheduled for anestrus or weaning periods, not when bitches are in proestrus or are pregnant.

Pups that have nursed on colostrum during the first 12 hours after birth have received passive immunity to viruses against which the dam was immunized. If pups cannot nurse on colostrum, 16 ml of pooled serum administered subcutaneously has been shown to be a successful alternative (Bouchard et al., 1992). Maternally acquired immunity declines over time, and the rate of decline, although variable, depends on the level of the dam's immunity at parturition and the amount of colostrum ingested by each pup. About 30-50 percent of pups will be susceptible to disease and capable of being effectively vaccinated by the age of 6-7 weeks. Most pups (more than 95 percent) can be effectively vaccinated by the age of 16 weeks. General principles of immunity in newborn animals and of immunoprophylaxis are reviewed elsewhere (Carmichael, 1983; Tizard, 1977a,b). Diseases to which pups are susceptible and vaccination schedules are discussed in Chapter 5, Veterinary Care.

Roundworms (*Toxocara canis*) and hookworms (*Ancylostoma caninum* and *A. braziliense*) are endoparasites that commonly infect young pups. Roundworms are typically transmitted from bitches to pups in utero, and pups begin to shed eggs in their feces 3 weeks after birth. Pups infected with hookworm larvae in their dams' milk typically begin to pass eggs in their feces 2 weeks after birth. It is important that pups receive treatment early in life if infection with roundworms or hookworms is suspected. To prevent peracute hookworm disease in unweaned pups of bitches harboring large numbers of somatic larvae, it might be necessary to treat the pups before hookworm eggs are detectable in fecal examinations. Canine endoparasites are reviewed in Chapter 5 and discussed fully elsewhere (Georgi and Georgi, 1992).

SOCIALIZATION OF PUPS

There is ample evidence of the importance of adequate socialization for the normal behavioral development of dogs (Clarke et al., 1951; Fox, 1968; Freedman et al., 1961; Houpt, 1991; Scott and Fuller, 1965). The term *socialization* is somewhat confusing because it has been used to describe events, processes, and procedures. In the narrowest sense, socialization is the development of the primary social attachments that form between a pup, its dam, and its littermates during a critical or sensitive period in its behavioral development (Scott, 1968). The process is not peculiar to dogs but occurs in many species of social mammals (see, for example, Cairns, 1966; Harlow and Harlow, 1969). In a broader sense, socialization is the process by which pups form attachments to other dogs, people, and environments. Attachment formation might require nothing more than sufficient exposure to or experience with other dogs, people, and elements of the environment, which results in familiarity with a variety of stimuli (Cairns, 1966; Scott, 1963). Breeds and individual pups differ in ease of socialization (Scott, 1970). In any case, adequate socialization allows a pup to develop normal social relationships with other dogs and to adapt to pair or group housing, to adjust more easily to unfamiliar stimuli and environmental changes, and to accept handling with little or no fear and distress (Scott, 1980).

Sensitive Period for Socialization

There is a sensitive period for socialization during which attachments form most readily and rapidly (Scott and Fuller, 1965). The beginning of the period is marked by the startle response to sound at the age of approximately 3 weeks. Also at 3 weeks, a pup begins to display distress vocalizations when separated from its dam. Distress vocalizations are distinct from those made in response to fear (Davis et al., 1977), hunger (Compton and Scott, 1971; Scott and Bronson, 1964), or physical discomfort (Gurski et al., 1980). Separation distress is greater in an unfamiliar pen (Elliot and Scott, 1961). To minimize separation distress, pups should remain with their dams for at least their first 6 weeks.

Ease of attachment formation varies between breeds and individuals but generally peaks between the age of 6-8 weeks (Scott and Bronson, 1964). Although socialization probably occurs at a low rate throughout life, the end of the sensitive period is marked by the pup's increasing fear of the unfamiliar at the age of 12-14 weeks (Scott, 1962).

Consequences of Inadequate Socialization

Pups that are inadequately socialized during the sensitive period exhibit abnormal behaviors, called kennel-dog or isolation syndromes, that are characterized by one or more of the following behaviors: generalized fearfulness, fear-motivated aggression, timidity, immobility, or hyperactivity (Scott et al., 1967). Dogs that, as a result of inadequate socialization, become highly distressed when subjected to common laboratory procedures (e.g., handling, walking on a leash, restraint, venipuncture, moves to different enclosures, and contact with other dogs) probably do not make good research subjects and might be in a compromised state of well-being. It has been reported that physiologic measurements on such dogs can fall outside normal limits (Vanderlip et al., 1985b).

Socialization Programs

Providing contact and handling only during routine husbandry procedures might not be sufficient to produce behaviorally normal, cooperative research animals (Vanderlip et al., 1985a,b). Specific programs that address each aspect of socialization—to dogs, to people, and to the environment—should be implemented. Programs that can be used as examples for providing adequate socialization have been reported (Vanderlip et al., 1985a,b; Wolfle, 1990).

The following are examples of elements that might be included in socialization programs: positive contacts with more than one person, opportunities to follow handlers, introduction to some type of restraint (e.g., a collar and leash), contacts with conspecifics other than littermates, and opportunities to explore outside the kennel. Exploration might include exposure to floors of different textures, to a room with different lighting, to stairs, and to such equipment as exam tables, clippers, and scales. Exposures to those elements should be gradual and paired with positive reinforcers, such as food, petting, or verbal praise. Negative reinforcement and physical punishment can elicit aggressive or fearful behaviors and will make pups more difficult to handle. It is not necessary, or practical, to introduce pups to every type of environment, person, or animal to which they will later be exposed in order to provide adequate socialization. Evidence suggests that experience in coping successfully with change facilitates later success (Scott, 1980). Thus, the adequacy of any socialization program can be determined by the ability of pups to adapt successfully to environmental changes with minimal behavioral and physiologic disruption.

RECORD KEEPING

Records on colony reproduction are essential. Individual records should contain the following minimal information on each bitch:

* start date of each proestrus;
* dates of mating and stud dog's identification number;
* date on which bitch's diet should be increased (day 42 of gestation), date to move bitch to whelping facility (day 50), and range of expected whelping dates;
* actual whelping date, whelping complications, number and sex of live pups, number of stillbirths, and any obvious abnormalities in the pups;
* date to start weaning, bitch's distemper antibody titer (if known), and dates to deworm and vaccinate litter; and
* date(s) and details of disposition of litter.

In addition, missed cycles, abortions, or any abnormal maternal behavior should be recorded.

To facilitate review of the reproduction records of an entire colony, it is helpful to have a separate computerized or manual-entry spreadsheet that displays every reproductive cycle of each bitch in the colony. The spreadsheet is most useful if it lists the following information, organized chronologically by date of proestrus:

* identification number of each bitch whose proestrus was first observed on that date;
* for each bitch bred, identification number of stud dog, first and last dates of mating, total number of matings, and calculated or expected dates for medical examinations, moving to whelping facility, and whelping; and
* expected date of next cycle.

The spreadsheet should be updated periodically to include for each bitch the actual whelping date; the length of gestation; litter information, as described above; the actual date of the next cycle; and the calculated interestrus interval. A computerized list can be sorted to review the breeding records of individual bitches and males over several years. Such a list also allows examination for trends in low fertility, long or short gestation lengths as indicators of poorly timed inseminations, number of matings per cycle, projected periods during which several bitches will be in heat at the same time or no bitches will be in heat, and other matters that could reflect husbandry, management, or staff problems that need correction.

REFERENCES

AAFCO (Association of American Feed Control Officials), Canine Nutrition Expert Subcommittee, Pet Food Committee. 1993. AAFCO nutrient profiles for dog foods. Pp. 92-99 in Official Publication 1993. Atlanta: Association of American Feed Control Officials. Available from Charles P. Frank; AAFCO Treasurer; c/o Georgia Department of Agriculture; Plant Food, Feed, and Grain Division; Capitol Square, Atlanta, GA 30334.

Amann, R. 1986. Reproductive physiology and endocrinology of the dog. Pp. 532-538 in Current Therapy in Theriogenology 2. Diagnosis, Treatment and Prevention of Reproductive Diseases in Small and Large Animals, D. A. Morrow, ed. Philadelphia: W. B. Saunders.

Andersen, A. C. 1970. Reproduction. Pp. 31-39 in The Beagle as an Experimental Dog. Ames: Iowa State University Press.

Andersen, A. C., and M. E. Simpson. 1973. The Ovary and Reproductive Cycle of the Dog (Beagle). Los Altos, Calif.: Geron-X. 290 pp.

Bouchard, G. F., N. Solorzano, P. W. Concannon, R. S. Youngquist, and C. J. Bierschwal. 1991a. Determination of ovulation time in bitches based on teasing, vaginal cytology, and ELISA for progesterone. Theriogenology 35:603-611.

Bouchard, G., R. S. Youngquist, B. Clark, P. W. Concannon, and W. F. Braun. 1991b. Estrus induction in the bitch using a combination diethylstilbestrol and FSH-P. Theriogenology 36:51-65.

Bouchard, G., H. Plata-Madrid, R. S. Youngquist, G. M. Buening, V. K. Ganjam, G. F. Krause, G. K. Allen, and A. L. Paine. 1992. Absorption of an alternate source of immunoglobulin in pups. Am. J. Vet. Res. 53:230-233.

Breazile, J. E. 1978. Neurologic and behavioral development in the puppy. Vet. Clin. North Am. 8:31-45.

Burke, T. J., ed. 1986. Small Animal Reproduction and Infertility. Philadelphia: Lea & Febiger. 408 pp.

Cairns, R. B. 1966. Attachment behavior in mammals. Psychol. Rev. 73:409-429.

Carmichael, L. E. 1983. Immunization strategies in puppies—Why failures? Compend. Contin. Educ. Pract. Vet. 5:1043-1051.

Christiansen, I. J. 1984. Reproduction in the Dog and Cat. London: Balliere Tindall. 309 pp.

Clarke, R. S., W. Heron, M. L. Fetherstonhaugh, D. G. Forgays, and D. O. Hebb. 1951. Individual differences in dogs: Preliminary report on the effects of early experience. Can. J. Psychol. 5:150-156.

Compton, J. M., and J. P. Scott. 1971. Allelomimetic behavior system: Distress vocalization and social facilitation of feeding in Telomian dogs. J. Psychol. 78:165-179.

Concannon, P. W. 1991. Reproduction in the dog and cat. Pp. 517-554 in Reproduction in Domestic Animals, 4th ed., P. T. Cupps, ed. New York: Academic Press.

Concannon, P. W. 1992. Methods for rapid induction of fertile estrus in dogs. Pp. 960-963 in Current Veterinary Therapy. XI. Small Animal Practice, R. W. Kirk and J. D. Bonagura, eds. Philadelphia: W. B. Saunders.

Concannon, P. W., and M. Battista. 1989. Canine semen freezing and artificial insemination. Pp. 1247-1259 in Current Veterinary Therapy. X. Small Animal Practice, R. W. Kirk, ed. Philadelphia: W. B. Saunders.

Concannon, P. W., and G. B. DiGregorio. 1986. Canine vaginal cytology. Pp. 96-111 in Small Animal Reproduction and Infertility, T. Burke, ed. Philadelphia: Lea & Febiger.

Concannon, P. W., and D. H. Lein. 1989. Hormonal and clinical correlates of ovarian cycles, ovulation, pseudopregnancy, and pregnancy in dogs. Pp. 1269-1282 in Current Veterinary Therapy. X. Small Animal Practice, R. W. Kirk, ed. Philadelphia: W. B. Saunders.

Concannon, P., S. Whaley, D. Lein, and R. Wissler. 1983. Canine gestation length: Variation related to time of mating and fertile life of sperm. Am. J. Vet. Res. 44: 1819-1821.

Concannon, P. W., D. B. Morton, and B. J. Weir, eds. 1989. Dog and cat reproduction, contraception and artificial insemination. J. Reprod. Fert. Suppl. 39:1-350.

Davis, K. L., J. C. Gurski, and J. P. Scott. 1977. Interaction of separation distress with fear in infant dogs. Dev. Psychobiol. 10:203-212.

Elliot, O., and J. P. Scott. 1961. The development of emotional distress reactions to separation, in puppies. J. Genet. Psychol. 99:3-22.

Feldman, E. C., and R. W. Nelson. 1987. Canine and Feline Endocrinology and Reproduction. Philadelphia: W. B. Saunders. 564 pp.

Fox, M. W. 1968. Socialization, environmental factors, and abnormal behavioral development in animals. Pp. 332-355 in Abnormal Behavior in Animals, M. W. Fox, ed. Philadelphia: W. B. Saunders.

Freedman, D. G., J. A. King, and O. Elliot. 1961. Critical period in the social development of dogs. Science 133:1016-1017.

Georgi, J. R., and M. E. Georgi. 1992. Canine Clinical Parasitology. Philadelphia: Lea & Febiger. 227 pp.

Greco, D. S., and J. W. Watters. 1990. The physical examination and radiography. Pp. 1-17 in Veterinary Pediatrics: Dogs and Cats from Birth to Six Months, J. D. Hoskins, ed. Philadelphia: W. B. Saunders.

Gurski, J. C., K. Davis, and J. P. Scott. 1980. Interaction of separation discomfort with contact comfort and discomfort in the dog. Dev. Psychobiol. 13:463-467.

Harlow, H. F., and M. K. Harlow. 1969. Effect of various mother-infant relationships on rhesus monkey behaviors. Pp. 34-60 in Determinants of Infant Behavior IV, B. M. Foss, ed. London: Methuen.

Hedhammer, Å. 1981. Nutrition as it relates to skeletal diseases. Pp. 41-44 in Proceedings of the Kal Kan Symposium for the Treatment of Small Animal Diseases (Oct. 11-12, 1980), L. D. Howell, ed. Vernon, Calif.: Kal Kan Foods, Inc. Available from Kal Kan Foods, Inc., 3250 E 44th Street, Vernon, CA 90058-0853.

Hedhammer, Å., F. M. Wu, L. Krook, H. F. Schryver, A. Delahunta, J. P. Whalen, F. A. Kallfelz, E. A. Numez, H. F. Hintz, B. E. Sheffy, and G. D. Ryan. 1974. Overnutrition and skeletal disease. An experimental study in growing Great Dane dogs. Cornell Vet. 64(suppl. 5):1-159.

Hegstad, R. L., and S. D. Johnston. 1992. Use of serum progesterone ELISA tests in canine breeding management. Pp. 943-947 in Current Veterinary Therapy. XI. Small Animal Practice, R. W. Kirk and J. D. Bonagura, eds. Philadelphia: W. B. Saunders.

Holme, D. W. 1982. Practical use of prepared foods for dogs and cats. Pp. 47-59 in Dog and Cat Nutrition, A. T. B. Edney, ed. New York: Pergamon Press.

Holst, P. A. 1986. Vaginal cytology in the bitch. Pp. 457-462 in Current Therapy in Theriogenology 2. Diagnosis, Treatment and Prevention of Reproductive Diseases in Small and Large Animals, D. A. Morrow, ed. Philadelphia: W. B. Saunders.

Houpt, K. A. 1991. Domestic Animal Behavior for Veterinarians and Animal Scientists, 2d ed. Ames: Iowa University Press. 408 pp.

Johnson, C. A., ed. 1986. Reproduction and periparturient care. Vet. Clin. N. Am. 16(3):1-605.

Johnston, S. D. 1986. Parturition and dystocia in the bitch. Pp. 500-501 in Current Therapy in Theriogenology 2. Diagnosis, Treatment and Prevention of Reproductive Diseases in Small and Large Animals, D. A. Morrow, ed. Philadelphia: W. B. Saunders.

Johnston, S. D., and S. E. Romagnoli, eds. 1991. Canine Reproduction. Vet. Clin. N. Am. 21(3):421-640.

Jones, D. E., and J. O. Joshua. 1988. Reproductive Clinical Problems in the Dog, 2d ed. London: Wright. 238 pp.

Kealy, R. D., S. E. Olsson, K. L. Monti, D. F. Lawler, D. N. Biery, R. W. Helms, G. Lust, and G. K. Smith. 1992. Effects of limited food consumption on the incidence of hip dysplasia in growing dogs. J. Am. Vet. Med. Assoc. 201:857-863.

Lawler, D. F. 1989. Care and diseases of neonatal puppies and kittens. Pp. 1325-1333 in Current Veterinary Therapy. X. Small Animal Practice, R. W. Kirk, ed. Philadelphia: W. B. Saunders.

Lewis, L. D., M. L. Morris, Jr., and M. S. Hand. 1987. Dogs—Feeding and Care Pp. 3.1-3.32 in Small Animal Clinical Nutrition III. Topeka, Kans.: Mark Morris Associates. Available from Mark Morris Associates, 5500 SW 7th Street, Topeka, KS 66606.

Morton, D. B., and S. G. Bruce. 1989. Semen evaluation, cryopreservation and factors relevant to the use of frozen semen in dogs. J. Reprod. Fert. Suppl. 39:311-316.

NRC (National Research Council), Board on Agriculture, Subcommittee on Dog Nutrition, Committee on Animal Nutrition. 1985. Nutrient requirements and signs of deficiency. Pp. 2-38 in Nutrient Requirements of Dogs, revised ed. Washington, D.C.: National Academy Press.

Olson, P. N., M. A. Thrall, P. M. Wykes, P. W. Husted, T. M. Nett, and H. R. Sawyer, Jr. 1984. Vaginal cytology. I. A useful tool for staging the canine estrous cycle. Compend. Contin. Educ. Pract. Vet. 6:288-298.

Patterson, D. F. 1975. Diseases due to single mutant genes. J. Am. Anim. Hosp. Assoc. 11:327-341.

Patterson, D. F., G. A. Aguirre, J. C. Fyfe, U. Giger, P. L. Green, M. E. Haskins, P. F. Jezyk, and V. N. Meyers-Wallen. 1989. Is this a genetic disease? J. Small Anim. Pract. 30:127-139.

Poffenbarger, E. M., M. L. Chandler, S. L. Ralston, and P. N. Olson. 1990. Canine neonatology. Part 1. Physiologic differences between puppies and adults. Compend. Cont. Educ. Pract. Vet. 12:1601-1609.

Scott, J. P. 1962. Critical periods in behavioral development. Science 138:949-958.

Scott, J. P. 1963. The process of primary socialization in canine and human infants. Soc. Res. Child Dev. Monogr. 28(1):1-47.

Scott, J. P. 1968. The process of primary socialization in the dog. Pp. 412-439 in Early Experience and Behavior, G. Newton and S. Levine, eds. Springfield, Ill.: Charles C Thomas.

Scott, J. P. 1970. Critical periods for the development of social behaviour in dogs. Pp. 21-32 in The Post-Natal Development of Phenotype, S. Kazda and V. H. Denenberg, eds. Prague: Academia.

Scott, J. P. 1980. The domestic dog: A case of multiple identities. Pp. 129-143 in Species Identity and Attachment: A Phylogenetic Evaluation, M. A. Roy, ed. New York: Garland STPM.

Scott, J. P., and F. H. Bronson. 1964. Experimental exploration of the et-epimeletic or care-soliciting behavioral system. Pp. 174-193 in Psychobiological Approaches to Social

Behavior, P. H. Leiderman and D. Shapiro, eds. Stanford, Calif.: Stanford University Press.

Scott, J. P., and J. L. Fuller. 1965. Genetics and the Social Behavior of the Dog. Chicago: University of Chicago Press. 468 pp.

Scott, J. P., J. H. Shepard, and J. Werboff. 1967. Inhibitory training of dogs: Effects of age at training in basenjiis and Shetland sheepdogs. J. Psychol. 66:237-252.

Shille, V. M. 1986. Management of reproductive disorders in the bitch and queen. Pp. 1225-1229 in Current Veteterinary Therapapy. IX. Small Animal Practice, R. W. Kirk, ed. Philadelphia: W. B. Saunders.

Shille, V. M., M. B. Calderwood-Mays, and M.-J. Thatcher. 1984. Infertility in a bitch associated with short interestrous intervals and cystic follicles: A case report. J. Am. Anim. Hosp. Assoc. 20:171-176.

Shultz, F. T. 1970. Genetics. Pp. 489-509 in The Beagle as an Experimental Dog, A. C. Andersen, ed. Ames: Iowa State University Press.

Tizard, I. R. 1977a. Immunity in the fetus and newborn animal. Pp. 155-168 in An Introduction to Veterinary Immunology. Philadelphia: W. B. Saunders.

Tizard, I. R. 1977b. Immunoprophylaxis: General principles of vaccination and vaccines. Pp. 169-183 in An Introduction to Veterinary Immunology. Philadelphia: W. B. Saunders.

Vanderlip, S. L., J. E. Vanderlip, and S. Myles. 1985a. A socializing program for laboratory-raised canines. Lab Anim. 14(1):33-36.

Vanderlip, S. L., J. E. Vanderlip, and S. Myles. 1985b. A socializing program for laboratory-raised canines. Part 2: The puppy socialization schedule. Lab Anim. 14(2):27-36.

Willis, M. B. 1989. Genetics of the Dog. London: H. F. & G. Witherby. 417 pp.

Wolfle, T. L. 1990. Policy, program, and people: The three P's to well-being. Pp. 41-47 in Canine Research Environment, J. A. Mench and L. Krulisch, eds. Bethesda, Md.: Scientists Center for Animal Welfare. Available from SCAW, 4805 St. Elmo Avenue, Bethesda, MD 20814.

Yeager, A. E., and P. W. Concannon. 1990. Association between the preovulatory luteinizing hormone surge and the early ultrasonographic detection of pregnancy and fetal heartbeats in beagle dogs. Theriogenology 34:655-665.

5

Veterinary Care

Veterinary care in laboratory animal facilities goes beyond the prevention, diagnosis, treatment, and control of disease. It also includes monitoring animal care and welfare and providing guidance to investigators on handling and immobilizing animals and preventing or reducing their pain and distress (NRC, 1985, 1992). Responsibilities of the attending veterinarian are specified by the Animal Welfare Regulations (9 CFR 2.33, research facilities; 9 CFR 2.40, dealers and exhibitors).

The first sections of this chapter deal with the procurement and conditioning of research dogs and the control of infectious and parasitic diseases. Aspects of veterinary care dealing with the use of anesthetics and analgesics, surgery and postsurgical care, and euthanasia are taken up in the last three sections. The medical aspects of reproductive disorders are discussed in Chapter 4; special care for pups is also reviewed in Chapter 4 and addressed in detail elsewhere (Hoskins, 1990). Reference values for blood analytes can be found in textbooks by Kaneko (1989) and Loeb and Quimby (1989).

Dogs can be afflicted with many uncommonly occurring but scientifically interesting diseases and disorders, many of which also afflict humans. Some breeds have predispositions to particular diseases and disorders (e.g., dalmatians are prone to urate bladder stones); a comprehensive review of this subject is available (Willis, 1989). Chapter 6 of this book addresses the maintenance of dogs with selected genetic disorders.

PROCUREMENT

General Considerations

Dogs acquired from outside a research facility's breeding program must be obtained lawfully from dealers licensed by the U.S. Department of Agriculture (USDA) or sources that the USDA has exempted from licensing (7 USC 2137). A *List of Licensed Dealers* can be obtained from Regulatory Enforcement and Animal Care, Animal and Plant Health Inspection Service, USDA, Federal Building, Room 268, 6505 Belcrest Road, Hyattsville, MD 20782. Examples of exempt sources are municipal pounds and people who provide dogs without compensation.

Procurement of dogs for research requires planning by a knowledgeable person to ensure that the dogs receive good care and that the needs of the investigator are met. The person should be familiar with federal regulations applicable to the acquisition of dogs (9 CFR, parts 2 and 3) and with state and local ordinances applicable to the aquisition of dogs from pounds and shelters. It is strongly recommended that institutions inspect vendors' premises for compliance with procurement specifications agreed on by contract before the first dogs are purchased and periodically thereafter.

Sources

Both random-source and purpose-bred dogs can be purchased for research purposes. Random-source dogs are those raised under unknown conditions of breeding and health. Sometimes they are stabilized and conditioned (see below) by the dealer before sale. Purpose-bred dogs are those from known matings that have limited exposure to infectious diseases.

Random-source dogs that have not been stabilized and conditioned by the vendor (often called nonconditioned random-source dogs) are usually acquired from USDA-licensed dealers or, less commonly, from pounds. If a number of dogs of similar weight or body conformation are needed, the purchaser must allow sufficient time for the group to be assembled by the vendor. Random-source dogs that have been stabilized and conditioned (often called conditioned random-source dogs) should be purchased only from vendors that have written standard procedures for their conditioning programs. Purpose-bred dogs are acquired from USDA-licensed dealers that breed dogs specifically for research or from an institutional breeding program. Dogs with diseases of research interest are often acquired from exempt sources, such as pet owners referred by clinical veterinarians.

Conditioning

Conditioning is defined as physiologic and behavioral adjustment to a new environment. The period required for that adjustment to occur is called the conditioning period. Conditioning consists of adjustment to a new regimen, including new people, diet, climate, and exercise. The adjustment can be hastened if the using institution provides the same type of food as the dealer or vendor and uses the same type of automatic watering devices. Physiologic status, as well as the presence of diseases, can be determined by assessing red-cell counts, packed-cell volumes, and white-cell differential counts and by using blood urea nitrogen tests and other examinations of blood and urine. Those tests are most valuable when samples are taken several days after arrival, by which time initial adjustments to the new environment have been made. Abnormal findings on any of the tests might warrant followup examinations.

Evidence of behavioral adjustment includes decreases in fearful behaviors, increases in friendly behaviors, increases in playfulness, and normal grooming behaviors. Some dogs might not adapt to human handling or the environment and are therefore inappropriate for use in long-term studies. The dealer should be questioned about the sources and histories of such dogs to determine whether additional dogs purchased from that dealer will be similarly distressed. Information on maladaptive canine behavior has been published elsewhere (Scott, 1970).

Many procedures—such as trimming of nails, removal of matted fur, bathing, and teeth-cleaning—can be performed during the conditioning period.

There is no definitive rule about the optimal period for conditioning. The intended use of the dogs, the season, prevalence of canine diseases in the area, and other factors influence the length of the conditioning period. If the dogs are well selected, adequately socialized, immunized, and treated for parasites before delivery, the conditioning period can be reduced. Random-source dogs that have been held for 10 days or more by the dealer usually require at least 21 days of conditioning at the institution before one can be confident that they have adapted fully. Some prefer a minimal conditioning period of 45 days. The importance of humane treatment and proper care during conditioning must be emphasized.

CONTROL OF INFECTIOUS DISEASES

General Considerations

There are three important strategies for controlling canine infectious diseases: examining dogs on arrival and refusing to accept dogs that exhibit signs of disease, placing all newly acquired dogs in quarantine, and isolating dogs that become sick. Some infectious pathogens to which dogs are susceptible could be introduced into an established colony by new arrivals, especially by random-source dogs, which are commonly unvaccinated. The most common of these pathogens are canine distemper virus (CDV); canine parvovirus (CPV-2); canine herpesvirus (CHV); the respiratory agents canine parainfluenza (PI-2), *Bordetella bronchiseptica*, and *Mycoplasma* spp.; canine adenovirus type 1 (CAV-1, infectious canine hepatitis) and type 2 (CAV-2, tracheobronchitis virus); and canine coronavirus (CCV). An additional problem that warrants careful consideration is the possibility that unvaccinated random-source dogs can harbor rabies virus, which can have a long incubation period (Acha and Szyfres, 1987). Dermatophytosis (ringworm, principally *Microsporum canis* and *M. gypseum*) and canine papillomatosis (warts) can also present problems. Protection against these pathogens is discussed briefly below. Detailed information on canine infectious diseases is available in a number of general references (e.g., Appel, 1987; Barlough, 1988; Greene, 1990).

Quarantine

Quarantine (in this context, the isolation of newly acquired animals until their health status has been evaluated) minimizes the risk of spreading diseases from newly arrived dogs to those already in the colony. In most facilities, the quarantine and conditioning periods overlap. During the quarantine period, most attention is directed to the control of infectious diseases and parasites. Procurement of dogs that are free of infectious diseases and parasites (i.e., conditioned random-source dogs or dogs bred specifically for research) reduces the time necessary for both quarantine and conditioning and might result in more reliable research results. Nonconditioned random-source dogs should be quarantined as a group, and no additional nonconditioned dogs should be introduced into the group.

Quarantine facilities should be designed to provide physical barriers to the spread of infectious diseases (e.g., unidirectional airflow). That is especially important when the research and quarantine facilities are parts of a single building. It is preferable for a quarantine facility to have its own animal-care technicians; however, if this is not possible, quarantined dogs should be cared for last.

Newly arrived dogs should be housed singly to enable veterinarians and technicians to determine which dogs are not eating well, exhibit signs of disease, or are abnormal in other ways. Ideally, dogs are vaccinated by the dealer. If not, they should be vaccinated as soon as possible after arrival against CDV, CPV-2, and CAV-2 (such vaccination also protects against

CAV-1). Vaccination for PI-2 and *B. bronchiseptica* should be considered in institutions where respiratory disease is common. If the dogs are to be vaccinated against leptospirosis and rabies, that is usually done at the same time. If a person is bitten or scratched, the injured area should be cleansed, the person should be referred to appropriate medical personnel, and the dog should be isolated for at least 10 days, as recommended by the National Association of State Public Health Veterinarians (1993).

Research and Breeding Colonies

The major threat to an established colony is that newly introduced dogs might harbor an infectious-disease agent or that personnel might carry such an agent into the colony on their hands or clothing. A regular immunization program, quarantine of nonconditioned random-source dogs, and rigorous sanitation practices will help to protect against infectious agents inadvertently introduced into an established colony. Annual vaccination with a multivalent vaccine is generally recommended, although immunity to CDV and CPV-2 generally persists for at least 3 years. In areas in which respiratory disease is common, frequent vaccination (every 3 months) might be indicated. Frequent vaccination (every 6 months) with leptospira bacterins is recommended in areas in which leptospirosis is endemic or is a proven problem.

Breeding Stock

Some infectious diseases are of special concern in breeding colonies. CHV can remain undetected in a breeding kennel for years. When susceptible, pregnant (usually young) bitches are introduced into the colony, latent CHV manifests itself by causing abortions or fetal or neonatal deaths. A detailed discussion of CHV is available elsewhere (Carmichael and Greene, 1990a).

Canine brucellosis can severely affect reproduction in a breeding kennel. It is also a zoonosis. All dogs purchased for breeding stock should be tested for *Brucella canis* antibodies on arrival and placed in quarantine for at least 1 month, at which time a second brucellosis test should be run. New dogs should not be introduced into a breeding colony unless both tests are negative. An infected dog should not be used for breeding or for long-term studies. Beagles have an unusually high prevalence of brucellosis, although it is occasionally diagnosed in random-source dogs (Carmichael, 1979). For detailed discussions of canine brucellosis see Carmichael (1990) and Carmichael and Greene (1990b).

Pups

CDV and CPV-2 infections are the principal viral diseases that threaten pups during the first 4 months of life, and prevention of these diseases should be the principal objective of an immunization program. Maternal antibody to CDV interferes with the development of an immune response to CDV vaccine; measles vaccine protects against disease but not infection in pups in which maternal antibody is still present (pups about 6-10 weeks old) (Baker, 1970). CDV vaccine should be given to dogs by 14 weeks of age. No vaccines can prevent parvovirus infection in pups during one critical period—that during which they still have maternal antibodies that inhibit the response to vaccination but do not protect against virulent CPV-2 (Carmichael, 1983). Proper management practices are critical in preventing this infection. If a pup does contract the disease, it should be isolated immediately, and rigorous disinfection procedures should be implemented. Diseases caused by adenoviruses and CCV can occur in pups, but they are less common.

It is generally recommended that modified live-virus vaccines be used for immunization, if available. A killed-virus vaccine is used for rabies. A multivalent vaccine that protects against distemper, hepatitis, leptospirosis, and parvovirus and parainfluenza infections can be used. An intranasal vaccine against *Bordetella bronchiseptica*, which causes kennel cough, is generally recommended. Several vaccination regimens have been proposed (Baker et al., 1961; Carmichael, 1983; Swango, 1983); one of them is given in Table 5.1 as a guide, but others are acceptable. The vaccination schedule should be adapted to address the perceived risk of infection.

Pups can be vaccinated with intranasal vaccine against *B. bronchiseptica* at 3-4 weeks of age. Other than that, vaccinating pups less than 6 weeks old is not recommended, because vaccine safety has not been studied in very young pups. Isolation is more important than vaccination in preventing disease in such pups.

TABLE 5.1   A Vaccination Schedule for Pups

Age	Vaccine
6 weeks	CDV or CDV combined with measles and CPV-2
8-10 weeks	CPV-2
12-14 weeks	Multivalent vaccine
16 weeks	CPV-2 (or multivalent vaccine) and rabies

Specific-Pathogen-Free Colonies

Dogs from known matings that have never been exposed to specific infectious agents are called specific-pathogen-free (SPF) for those agents. These dogs are used in infectious-disease and vaccine-development research in which animals are required not only to be free from pathogenic agents, but never to have been exposed, either naturally or through vaccination, to pathogenic agents. It might also be preferable to use SPF dogs in some transplantation studies, because in profound immunosuppression, native and vaccinal viruses (e.g., CDV and CAV-1) might be activated and cause disease (Thomas and Ferrebee, 1961).

The objective in preventing the outbreak of disease in SPF colonies is to isolate, rather than immunize, the dogs. Disease prevention depends on the establishment of physical barriers to preclude the introduction of disease agents, rigorous management practices, and control of personnel movement into and within the facility (Sheffy et al., 1961). Rodents and other pests that can transmit disease mechanically must be excluded. Purpose-bred SPF dogs are available commercially. If bred by the institution, initial breeding stock should be procured from dogs free of latent infectious agents, and all offspring taken by hysterectomy or cesarean section. Embryo-transfer technology offers additional possibilities for SPF colonies. Population immune status should be assessed periodically (at least once a year) by monitoring for antibodies to the common infectious diseases. In the event of an inadvertent infection that would compromise the use of the animals, the colony should be depopulated and re-established.

CONTROL OF PARASITIC DISEASES

Parasites are common in dogs, particularly random-source dogs. They can be found on the skin and hair and in the ears (ectoparasites) and in many internal organs, including the digestive tract, heart, lung, and blood vessels (endoparasites). Specific canine parasites are discussed briefly below; details on life cycles of, treatment for, and prevention or control of these parasites are found elsewhere (Georgi and Georgi, 1992).

Ectoparasites

Ectoparasites include ticks, mites, lice, and fleas. Most can be easily eradicated with insecticides. Three ectoparasites commonly carried by random-source dogs can pose problems if they are not eliminated during quarantine.

- The most damaging is probably the *Rhipicephalus sanguineus* tick. This tick can feed on dogs during all life-cycle stages, and once it enters a facility, it can be expensive to remove.
- Mange caused by *Sarcoptes scabei* is sometimes inadvertently introduced into a facility on a dog that shows no overt signs of dermatosis. This parasite can be a particular problem in dogs that are group-housed or housed in cages or runs that allow the touching of body parts among animals (e.g., through wire-mesh walls). Sarcoptic mange is treated by dipping the affected dogs and all dogs in contact with them in insecticide. It is probably also worthwhile to steam-clean enclosures and floors.
- Fleas are commonly brought into facilities by random-source dogs. The flea life cycle can be disrupted by cleaning enclosures daily to remove developing eggs and larvae. Another strategy is to house dogs in enclosures raised more than 33 cm above the floor. Fleas cannot jump higher than 33 cm, so fleas that fall to or develop on the floor cannot reach the dog to feed.

Additional ectoparasitic infestations that might persist in kennel settings include infestation with ear mites (*Otodectes cynotis*), "walking dandruff" (*Cheyletiella yasguri*), and lice (*Linognathus setosus, Trichodectes canis*, and *Heterodoxus spiniger*). The canine nasal mite (*Pneumonyssoides caninum*) can also persist, but it is not known how often infestations with this mite occur in random-source dogs.

To prevent the introduction of skin-dwelling ectoparasites, random-source dogs should be bathed or dipped before they are moved to the housing facility. Their ears should be examined and, if appropriate, treated for ear mites. Mites should be considered as the cause of persistant skin lesions, and appropriate action should be taken to make a correct diagnosis.

All dogs, including random-source and breeding-colony dogs, are probably host to the hair-follicle mite *Demodex canis*. Dogs probably become infected as puppies while nursing. Typically, the infestation is nonpathogenic; in rare instances, the mite causes severe mange. The development of demodectic mange in large numbers of kennel dogs is rare but has occurred. Treatment with topical applications or dips is possible as long as the lesions remain focal, but generalized demodectic mange often indicates some underlying problem (e.g., an inherited susceptibility to demodectic mange or a compromised immune system), and its treatment is difficult or impossible.

Endoparasites

SPF dogs and purpose-bred dogs often host both protozoan and helminthic endoparasites. The protozoa include *Isospora* spp., *Giardia* spp., trichomonads, *Cryptosporidium* spp., *Balantidium* spp., and amebas. The helminths include ascarids (e.g., *Toxocara canis* and *Toxascaris leonina*), *Filaroides* spp., *Strongyloides stercoralis*, and occasionally hookworms and whipworms. If dogs are housed in a manner that allows mosquitoes access to them, they are also susceptible to infection with heartworm, *Dirofilaria immitis*.

Isospora spp. have direct life cycles (i.e., no intermediate host is required). Oocysts of these coccidia are commonly present in the feces of young dogs raised in colonies, and more than one species can be present in one dog. The oocysts of *I. canis*, *I. ohioensis*, *I. neorivoltos*, and *I. burrowsi* are morphologically similar; however, those of *I. canis* are larger than those of the other three. Clinical signs include increased temperature and diarrhea that is occasionally bloody. Infections usually subside after several days to weeks. Oocyst shedding decreases to low numbers 4 weeks after it begins. Chemoprophylaxis and basic sanitation are necessary to control the infection if it causes problems.

Cryptosporidium is occasionally present in dogs in closed colonies, although it typically does not cause disease. In immunocompetent dogs, the small oocysts of *Cryptosporidium* are shed in low numbers, if at all, for a limited period; however, in immunosuppressed or immunocompromised dogs, *Cryptosporidium* can cause fatal disease. There is no proven method of chemoprophylaxis or treatment, but routine sanitation procedures, accompanied by regular steam cleaning of areas that might be contaminated, will assist in reducing exposure to oocysts.

Giardia canis is commonly present in both purpose-bred and SPF dogs. The prevalence is high in pups and decreases with age. The organism is spread between dogs by the fecal-oral transmission of resistant cysts. Typically, pups are infected with *Giardia* and one or more species of *Isospora*; however, the infection usually causes little or no disease. As dogs mature, the number of organisms decreases. As with *Isospora*, chemoprophylaxis and basic sanitation are the most effective means of controlling *Giardia*.

Trichomonas canistomae is a commensal organism present in the mouths of many dogs. It has no cyst stage and is transmitted between dogs by direct oral contact. There is usually no need for treatment. Species of *Trichomonas* and *Pentatrichomonas* are present in the large intestines of many laboratory-reared dogs. None of these species has a cyst stage; transmission is by the fecal-oral route. These organisms are sometimes observed in diarrheic feces in very large numbers, but they are usually not the cause of the diarrhea. Treatment is available but usually not necessary.

Balantidium coli, a large ciliated parasite that is rarely found in dogs, and *Entamoeba coli* and *E. histolytica*, smaller ameboid parasites, are transmitted by cysts passed in the feces. These parasites are present in the large bowel. Their life cycles are similar to that of *Giardia*, and once they are established in a colony, they are easily perpetuated.

The ascaridoid nematode (roundworm), *Toxocara canis*, is a common parasite of the small intestines of dogs, even in closed breeding colonies.

The parasite is transmitted from bitches to pups in utero, and pups begin to shed eggs in their feces a few weeks after birth. Once the eggs enter the environment, they require about 2 weeks to become infectious; they are very resistant to environmental extremes of heat, cold, and humidity. Pups should be treated soon after birth and several times during early life to prevent the development of adult roundworms from the stages obtained prenatally. Control measures should include steam cleaning of floors and disinfection of floors with a 1:4 (20 percent) solution of chlorine bleach. Adult dogs can have larvae in their tissues whether or not they are shedding eggs in their feces. It is possible to determine whether a dog has ever been infected by measuring antibody concentrations, and dogs that are *Toxocara canis*-naive are available commercially.

The other canine ascaridoid, *Toxascaris leonina*, has a direct life cycle and does not infect pups transplacentally. The eggs of this parasite develop more rapidly than those of *Toxocara canis* but are just as resistant to extremes of heat, cold, and humidity. *Toxascaris leonina* is commonly present in the small intestines of older purpose-bred and SPF dogs, but it is not known how the cycle is maintained in these colonies. Control and treatment are the same as those used for *Toxocara canis*.

Filaroides hirthi is present in the lung parenchyma of many purpose-bred and SPF dogs. The lung lesions caused by the parasite can confuse histopathologic evaluations in toxicologic experiments. The life cycle is direct, and infective larvae are transmitted between dogs by oral or fecal-oral contact. Immunosuppressed dogs can become seriously ill as a result of auto-reinfection that leads to heavy parasite burdens. Infections can be treated, but control is difficult because fecal assays are insensitive. Therefore, all dogs in a contaminated room must be treated, not just those with positive fecal tests. Proper sanitation is helpful, but the larvae do not persist for long periods in the environment.

Strongyloides stercoralis lives as a parthenogenetic female in the mucosa of the canine small intestine. Larvae develop to the infective stage 4-5 days after they are passed in feces. Transmission is by penetration of the skin by infective-stage larvae and by passage of tissue-dwelling larvae in the milk of lactating bitches. Immunosuppressed or immunocompromised dogs can develop severe disease as a result of auto-reinfection. *S. stercoralis* is also transmissible to humans. Although treatment is available, elimination of the parasite from a breeding colony is difficult because it is not certain that transmammary transmission can be interrupted by chemotherapeutic measures. Routine removal of feces and cleaning of cage or pen floors reduce transmission.

Adult hookworms live in the small intestine, where they cause blood loss and anemia. The hookworms *Ancylostoma caninum* and *A. braziliense*, like *S. stercoralis*, are transmitted through the milk or by larval penetration of the skin. However, infective-stage larvae are more likely to develop in soil than on a moist cage bottom fouled with feces, and transmission is more likely when dogs are housed outside on such surfaces as gravel or sand. Unlike dogs infected with *S. stercoralis*, dogs infected with hookworms often show signs of overt disease, characterized by bloody diarrhea. In addition, hookworm eggs are much easier to detect in feces than are *S. stercoralis* larvae. Those differences and the dissimilarity of conditions required for larval development make it much less likely that hookworms will persist undetected in a colony. The hookworm *Uncinaria stenocephala*, which is present in more temperate climates, is transmitted mainly by larval ingestion; skin penetration and transmission in milk are uncommon. Thus, *U. stenocephala* is less likely to be perpetuated in a closed colony.

Whipworms, *Trichuris vulpis*, live in the cecums and colons of dogs and cause large bowel disease that can produce bloody stools. The life cycle of this parasite is direct. Eggs are passed in feces and take several weeks to become infectious. They are highly resistant to environmental extremes, so contamination is very peristent if eggs get into the soil of earthen-floored runs. Dogs become infected by ingesting the infective eggs on soil-contaminated items. In the dog, the worms take about 3 months to develop to the adult stage, and reinfection is common. Treatment is available but often has to be repeated.

The filarioid nematode *Dirofilaria immitis* causes heartworm disease. It is transmitted between dogs by the bite of a mosquito. The prepatent period (the time between the inoculation of maturing forms by the mosquito and the first appearance of microfilariae in the host's blood) is slightly more than 6 months. The infection is often manifested as cardiopulmonary disease accompanied by respiratory distress and right-sided heart enlargement. In dogs with patent disease, infections can be diagnosed by demonstrating microfilariae in the blood; however, some infected dogs do not have circulating microfilariae (Glickman et al., 1984.). When it is important to ascertain that dogs are heartworm-free, serum or plasma can be examined with antigen-detection tests. Treatment for heartworm infection is generally precluded by its high cost, the stress it causes the dog, the length of time necessary for recovery, and the possibility of residual pathologic changes in the cardiovascular system.

Where *D. immitis* is enzootic, dogs given access to outside runs should be protected by chemical prophylaxis. If dogs cannot be placed on chemical prophylaxis, because of a study design or for other reasons, they can be protected by enclosing the outside kennels with screening.

In addition to infection with the same parasites found in purpose-bred and SPF dogs, random-source dogs are likely to be infected with parasites that are relatively rare or that require intermediate hosts as part of their life cycles. If the intermediate hosts are uncommon (e.g., snails, then crayfish for the lung fluke *Paragonimus kellicotti*), there is little chance that the infection will be maintained in a kennel. However, if the intermediate host is commonly present around dogs (e.g., fleas for the tapeworm *Dipylidium caninum*), the parasite will probably persist in the facility as long as the intermediate host is present. Additional parasites that can be found in random-source dogs include the tapeworm *Taenia* spp. (intermediate hosts, mammals), the intestinal fluke *Alaria canis* (snails, then frogs), the esophageal nematode *Spirocerca lupi* (beetles), and the stomach nematode *Physaloptera* spp. (beetles).

Two parasites that are found rarely in random-source dogs, *Echinococcus* spp. and *Trypanosoma cruzi*, are important because they cause zoonoses. Larval stages of the canine tapeworms *Echinococcus granulosus* and *E. multilocularis* can be transmitted to humans in contaminated feces and cause unilocular and multilocular hydatid disease, respectively. Eggs of *Echinococcus* spp. are infectious when passed in feces and cannot be distinguished morphologically from eggs of taeniid tapeworms. *E. granulosus* is present in focal areas of the United States; *E. multilocularis* is present in the far northern continental United States, Alaska, and Canada. *Trypanosoma cruzi*, which is present in the southern United States, is a hemoflagellated protozoan that can infect the blood and tissues of opossums, armadillos, dogs, humans, and other mammals. Humans are infected by accidental self-inoculation with blood products from an infected animal. People handling dogs from areas where *Echinococcus* spp. and *T. cruzi* are enzootic should be made aware that such infections, although rare, are possible and can be associated with life-threatening conditions in humans.

Three other uncommon canine pathogens, all requiring arthropod vectors, have occasionally been diagnosed in dog facilities: *Leishmania* spp., *Babesia* spp., and *Ehrlichia canis*. The clinical signs caused by these pathogens are often poorly deliniated, so they can be harder to diagnose than common helminth infections.

Cutaneous and visceral leishmaniasis, caused by infections with various species of *Leishmania*, have been reported in both kennels and research colonies. The organism is typically transmitted between dogs by the bite of a phlebotomine sandfly, although the mode of transmission in the reported cases is not certain. Diagnosis is typically made by identifying the organisms histopathologically or serologically. Treatment is difficult but possible.

Babesiosis, caused by *Babesia canis* or *Babesia gibsoni*, can be introduced into colonies or kennels through an infected dog, an infected tick, or a blood transfusion. Once it is in an establishment, horizontal transmission typically occurs through exposure to infected blood that is not handled properly or through ticks, particularly *Rhipicephalus sanguineus*. Dogs with babesiosis display regenerative anemia, i.e., the bone marrow remains functional, and increased numbers of immature erythrocytes appear in the blood. The disease can be diagnosed by demonstrating the organisms in erythrocytes on stained blood films. Treatment is difficult, and drugs routinely used in parts of the world where babesiosis is common are not easily obtained in the United States.

Transmission of ehrlichiosis, a rickettsial infection caused by *Ehrlichia canis*, is similar to transmission of babesiosis. Signs of ehrlichiosis in dogs include fever, anorexia, epistaxis (nosebleeds), and reduced kidney function. Diagnosis is made serologically or by demonstrating the presence of the organism in blood smears. Treatment can alter the course of the disease but does not prevent an affected dog from becoming a carrier of the infection.

Good sanitation is probably the major means for controlling endoparasites in a dog facility. In facilities that house purpose-bred or SPF dogs, feces from healthy animals of different ages should be examined periodically for subclinical helminth or protozoan infections. Fecal and blood examinations can be used to screen random-source dogs for parasites on arrival at the facility. To prevent the introduction of helminth parasites into a facility, random-source dogs might be treated for some infections with an anthelminthic. A practical choice would be a broad-spectrum anthelminthic that is active against both nematodes and tapeworms.

RECOGNITION AND ALLEVIATION OF PAIN AND DISTRESS

Recognition of Distress Induced by Pain

Distress can be defined as "an aversive state in which an animal is unable to adapt completely to stressors and the resulting stress . . ." (NRC, 1992, p. 4). Scientists have legal, ethical, and humane obligations to minimize distress in experimental animals. Moreover, there is a pragmatic reason to minimize distress. Unless a stressor (such as pain) is the subject of the experiment, distressed animals might provide erroneous data (Amyx, 1987). Pain is an important cause of distress and is usually produced by disease, injury, or surgery.

Table 5.2 lists some of the signs of pain in dogs. Dogs usually respond to acute pain by vocalizing and by protecting or guarding the area of perceived pain. Signs include withdrawing, attempting to bite if touched, and adopting unusual postures (e.g., the laterally flexed position commonly adopted after lateral thoracotomy). Low-grade pain can produce restlessness. Severe pain, especially if chronic, usually makes dogs appear depressed and lethargic. The decrease in activity can be accompanied by one or more of the following: shivering, inappetence, panting, howling, or whining.

The *U.S. Government Principles for Utilization and Care of Vertebrate*

TABLE 5.2   Signs of Pain in Dogs[a]

Sign	Comment
Guarding	Attempting to protect or move painful part away (e.g., hunched position after celiotomy or laterally flexed position after lateral thoracotomy), attempting to bite
Vocalization	Whining or whimpering when touched or forced to use affected part
Mutilation	Licking, biting, scratching, shaking, or rubbing affected part
Restlessness	Pacing, lying down and getting up, or shifting weight
Recumbency	For unusual length of time
Depression	Inappetence, reluctance to move, or difficulty in rising
Pallor	Pale mucous membranes, probably a result of vasoconstriction caused by an increase in sympathetic tone

[a]Adapted from Soma, 1987; printed with permission of the author, the American Association for Laboratory Animal Science, and the Scientists Center for Animal Welfare.

Animals Used in Testing, Research, and Training (published in NRC, 1985) states that "unless the contrary is established, investigators should consider that procedures that cause pain or distress in human beings may cause pain or distress in other animals." This statement makes it clear that most surgical interventions must be accompanied by adequate anesthesia and suitable postoperative analgesia. Table 5.3 lists the degree and duration of pain that can be expected after surgery on various parts of a dog's body. Although pain thresholds are similar between individuals and even between species, pain tolerance varies widely. Therefore, each dog should be observed and treated as an individual in determining the need to administer analgesics.

Alleviation of Pain

Anesthetics

General anesthesia is the most important way of alleviating pain associated with surgery, and several textbooks contain detailed descriptions of acceptable techniques for inducing general anesthesia in dogs (Booth, 1988a; Hall and Clarke, 1991; Lumb and Jones, 1984; Muir and Hubbell, 1989; Short, 1987). Inhalant agents (e.g., isoflurane, methoxyflurane, and halothane) are often best for this purpose because they allow close regulation of the duration and depth of anesthesia and rapid and controlled reversibility. However, special equipment is required for administering them. Nitrous oxide is not a general anesthetic in dogs and should be used only as an adjunct to other, more potent anesthetics.

General anesthesia can also be provided with injectable drugs, such as barbiturates (e.g., thiamylal, thiopental, and pentobarbital), propofol, or Telazol

TABLE 5.3 Signs, Degree, and Length of Surgically Produced Pain[a]

Surgical Site	Signs of Pain	Degree of Pain	Length of Pain
Head, eye, ear, mouth	Attempts to rub or scratch; self-mutilation; shaking; reluctance to eat, drink, or swallow; reluctance to move	Moderate to high	Intermittent to continual
Rectal area	Rubbing, licking, biting, abnormal bowel movement or excretory behavior	Moderate to high	Intermittent to continual
Bones	Reluctance to move, lameness, abnormal posture, guarding, licking, self-mutilation	Moderate to high: upper part of axial skeleton (humerus, femur) especially painful	Intermittent
Abdomen	Abnormal posture (hunched), anorexia, guarding	Not obvious to moderate	Short
Thorax	Reluctance to move, respiratory changes (rapid, shallow), depression	Sternal approach, high; lateral approach, slight to moderate	Continual
Spine, cervical	Abnormal posture of head and neck, reluctance to move, abnormal gait- "walking on eggs"	Moderate to severe	Continual
Spine, thoracic or lumbar	Few signs, often moving immediately	Slight	Short

[a]Based on observations of dogs. Reprinted from NRC, 1992.

(a mixture of tiletamine and zolazepam). Each injectable drug has properties that determine its duration of action and the route by which it is best administered. Ketamine is used as an anesthetic but its effectiveness as an analgesic for visceral pain is disputed (Booth, 1988b; Hughes and Lang, 1983). It should be used in combination with another analgesic agent when visceral pain is expected. It can also induce seizure-like activity in dogs unless it is used in conjunction with another drug, such as diazepam, acepromazine, or xylazine. Chloralose and urethane are injectable anesthetics that have been used in some experiments; however, chloralose alone is a poor anesthetic that produces little analgesia unless it is combined with an opiate such as morphine (Rubal and Buchanan, 1986), or a short-acting anesthetic (Flecknell, 1987). Urethane is mutagenic and carcinogenic (Auerbach, 1967; Mirvish, 1968); it should be used with caution and only for nonsurvival surgery.

Neuromuscular blocking agents (e.g., succinylcholine, atracurium, curare, gallamine, pancuronium, and vecuronium) have no anesthetic or analgesic properties. They must not be used alone for surgical restraint, although they may be used in conjunction with anesthetic doses of general anesthetic drugs (NRC, 1985).

Local anesthetics (e.g., lidocaine, mepivacaine, and bupivacaine) act to disrupt nerve conduction temporarily. When applied around a nerve, they produce analgesia in the region served by that nerve. However, these drugs have no depressant effect on the brain; dogs undergoing procedures under local anesthesia usually must be restrained physically or chemically (e.g., with tranquilizers or sedatives). Specific techniques for regional anesthesia are described in several texts (Hall and Clarke, 1991; Lumb and Jones, 1984; Muir and Hubbell, 1989; Skarda, 1987; Soma, 1971). Local anesthetics alone are ordinarily used for only the most minor of surgical interventions; but they can be given either intrathecally or epidurally (usually via the lumbosacral space) to provide segmental anesthesia of caudal body parts sufficient for major surgery (e.g., celiotomy) (Skarda, 1987).

Analgesics

Opioid analgesics are compounds that act at specific opioid receptor sites in the central nervous system to produce analgesia. Table 5.4 lists some of these compounds. They are not general anesthetics, but can be used for surgery when combined with other appropriate drugs (NRC, 1992). Opioid analgesics (e.g., oxymorphone) can be injected epidurally to control postsurgical pain for extended periods with minimal systemic effects (Popilskis et al., 1991).

Opioid agonists have been combined with tranquilizers to produce so-called neuroleptanalgesic combinations (e.g., a mixture of fentanyl and droperidol known by the trade name Innovar-Vet and produced by Pitman Moore, Mundelein, Ill.). Such combinations are capable of producing a state that

TABLE 5.4 Opioid Analgesics Used in Dogs[a]

Drug	Dose (mg/kg)	Route[b]
Buprenorphine	0.01-0.2	IV, IM
Butorphanol	0.2-0.5	IV, IM
Fentanyl	0.04	IV, IM
Meperidine	2.0-6.0	IM
Morphine	0.5-1.0	SC
Oxymorphone	0.2-0.4	IV, IM

[a]Data from Harvey and Walberg, 1987.
[b]IV = intravenous; IM = intramuscular; SC = subcutaneous

sufficiently resembles general anesthesia to permit some surgical procedures (Muir and Hubbell, 1989; Soma and Shields, 1964). Xylazine, which is classified as a sedative, has analgesic properties because of its action on central alpha-2 receptor sites.

The nonsteroidal anti-inflammatory analgesics include acetaminophen, aspirin, flunixin, and ibuprofen. These drugs inhibit prostaglandin synthesis. They are ordinarily used to relieve the acute or chronic pain associated with inflammation and have little place in the management of severe or acute pain that is not associated with inflammation (NRC, 1992).

Recognition of Distress Not Induced by Pain

Signs of distress caused by stressors other than pain include changes in behavior (e.g., unexpected aggression), maladaptive behaviors (e.g., stereotypies), and physical changes (e.g., weight loss). Experienced and attentive animal caretakers are of the utmost importance in early recognition of signs of distress. Changes in biochemical measurements (e.g., plasma cortisol concentration) can also help in recognition of distress.

Alleviation of Distress Not Induced by Pain

Distress caused by stressors other than pain is often related to husbandry practices. Understanding and meeting dogs' social and physical needs will minimize or prevent such distress (NRC, 1992).

Phenothiazine tranquilizers, such as acepromazine (0.03-0.05 mg/kg intravenously or intramuscularly, 1.0-3.0 mg/kg by mouth), are useful as preanesthetic drugs because they make unruly animals more tractable, reduce the doses of anesthetic drugs necessary to maintain anesthesia, and make recovery from anesthesia smoother. However, they can have unpredictable effects and cause some animals to become excited rather than tranquil (Voith, 1984). The phenothiazines have minimal antianxiety effects, and they are not the drugs of choice for decreasing fearful reactions (Marder, 1991).

Alpha-2 agonists, such as xylazine (0.3-1.0 mg/kg intravenously, 0.5-2.0 mg/kg intramuscularly), have many of the advantages of the phenothiazines and are also good analgesics (Gleed, 1987). However, they can cause serious cardiovascular depression, hyperglycemia, and depressed thermoregulation, which can be reversed with yohimbine if necessary (Denhart, 1992).

Benzodiazepines, such as diazepam (0.1-0.5 mg/kg intravenously, 0.3-0.5 mg/kg intramuscularly) are often used as adjuncts to injectable anesthetic drugs, such as the barbiturates and ketamine, because they reduce the dose necessary to produce anesthesia and provide muscle relaxation (Gleed, 1987). Diazepam (Valium) is also used alone to treat seizures. Like the phenothiazines, the benzodiazepines have an excitatory effect on some animals. Because they are the drugs of choice for the treatment of fearful behaviors (Marder, 1991), especially fear of people (Hart, 1985), they can be useful in reducing distress in unsocialized dogs. However, the benzodiazepines must be used with care in dogs that display fear-motivated aggression. Decreasing the fear might make such dogs more likely to attack (Marder, 1991).

SURGERY AND POSTSURGICAL CARE

Surgery in dogs should be performed in accordance with the tenets in the *Guide* (NRC, 1985). The requirements for minor and nonsurvival surgical procedures are less stringent than those for major survival surgical procedures.

Personnel performing surgical procedures must be adequately trained. Facilities for performing surgical procedures should be available as outlined in the *Guide* (NRC, 1985). The successful practice of survival surgery requires strict adherence to aseptic surgical technique, as well as provision of adequate postoperative care and analgesia for the experimental subject. Aseptic techniques also have some value in major nonsurvival surgical procedures (Slattum et al., 1991). Generally, only healthy conditioned or purpose-bred dogs should be used for survival surgery. Familiarizing the dog with the laboratory environment can assist investigators in identifying intractable subjects and can be beneficial in decreasing postoperative stress.

Presurgical Preparation

Dogs should be surgically prepared by careful shaving to remove all hair from the surgical field. Shaving reduces contamination of the wound and avoids delays in healing that can occur if hair becomes matted in the incision. If a thermal cautery is to be used, an area should also be shaved for placement of a ground lead. Adherent grounding pads are available. The surgical field should be thoroughly cleaned with Betadine (povidone-iodine) or another appropriate surgical scrubbing material. Betadine sterile solution or other appropriate preparation should be applied to the entire field and allowed to dry. Underpadding used to absorb such solutions can be flammable and should be removed before surgery.

All surgical instruments and chronic instrumentation must be sterilized with steam (autoclaving) or gas (ethylene oxide with proper poststerilization aeration time). Cold chemical sterilization is appropriate for minor surgical procedures, but exposure time must be adequate, and the instruments must be thoroughly rinsed in sterile saline before they come into contact with body tissues. All items should be packaged for sterilization in such a way that they can be opened and positioned for use without compromising sterility. Investigators should follow standard surgical practices: donning surgical caps and masks, scrubbing, and donning surgical gowns and gloves. Sterile drapes should be positioned on the dog to define the surgical field. During the course of surgery, procedures for preserving sterility should be strictly followed.

Generally, dogs should be treated with the appropriate preanesthetic medications (e.g., tranquilizers and atropine) to provide a degree of sedation and facilitate handling. General anesthesia is reviewed in the section "Alleviation of Pain" (see pages 64-67); the type used depends on the type and duration of the surgical procedure. The adequacy of anesthesia can be assessed by the absence of the eyelid reflex and by the lack of withdrawal in response to painful stimuli (e.g., toe pinch). Insertion of a cuffed endotracheal tube will ensure patency of the respiratory tract.

The physiologic status of dogs under general anesthesia should be assessed by monitoring such parameters as pulse rate, systemic blood pressure, and respiratory rate. Electrocardiography can be used to monitor the status of the heart. A heating pad is useful for maintaining body temperature. If inhalant anesthetics are used, the anesthetized dog should be ventilated (tidal volume, 15-20 ml/kg; respiratory rate, 13-20 breaths/minute), and carbon dioxide should be monitored. Respiratory rate, tidal volume, and inspiratory-expiratory ratio can be adjusted to achieve acceptable end-tidal carbon dioxide (38-40 torr) and blood oxygen saturation greater than 90 percent.

An intravenous catheter should be placed in the cephalic vein to provide a continuous intravenous drip (e.g., of lactated Ringer's solution) for volume replacement and to ensure rapid access to the circulatory system. Depending on the situation, antibiotics can be administered through the catheter or intramuscularly. There is evidence that giving antibiotics during the 2 hours before surgery is more beneficial than giving them either during or after surgery (Classen, 1992).

Postsurgical Care

Appropriate analgesics should be administered for postoperative pain, as needed (see pp. 66-67 and NRC, 1992). Surgical wounds and sites of instrument entry into the body should be cleaned and treated daily (e.g., with 0.3 percent hydrogen peroxide or dilute Betadine solution). Topical antibiotics (e.g., bacitracin ointment) can be applied. Surgical dressings should be changed every day.

Basic biologic functions—including urination, defecation, and appetite—are good indicators of a dog's overall physical well-being. These are easy to observe and should be monitored regularly and often. Followup clinical examinations and laboratory tests can be used to identify specific problems. Appropriate supportive care should be provided as needed.

A commonly used experimental protocol involving major survival surgery in the dog is the implantation of instruments that allow physiologic measurements over a long period while the dog is conscious. The dog is particularly suitable for this type of protocol because of its size, its equable temperament, and the close parallelism of its physiologic functions with those of humans. Strict adherence to the recommendations above will minimize confounding effects.

EUTHANASIA

Euthanasia is a method of killing an animal rapidly and painlessly (NRC, 1985). It should be carried out by trained personnel following current guidelines established by the American Veterinary Medical Association (AVMA) Panel on Euthanasia (AVMA, 1993 et seq.; NRC, 1985) The method used must produce rapid unconsciousness and subsequent death without evidence of pain or distress, or the animal must be anesthetized before being killed (9 CFR 1.1). The method used should also be safe for attending personnel, be easy to perform, and cause death without producing changes in tissues that might interfere with necropsy evaluation. Methods of euthanasia recommended by the AVMA Panel on Euthanasia (AVMA, 1993) are discussed below.

Injection of Lethal Substances

Injection of a lethal substance is probably the most suitable method for euthanatizing laboratory dogs. It usually involves the intravenous injection of a large dose of a barbiturate anesthetic, such as pentobarbital (more than 100 mg/kg). The advantage of this method is that the animal is anesthetized within seconds and does not undergo the pain or distress that might be associated with later respiratory and cardiac arrest. In fact, cardiac arrest can be delayed for many minutes after the onset of anesthesia; therefore, cardiotoxins (e.g., large doses of dibucaine) are sometimes used to hasten death (Wallach et al., 1981). Unruly or aggressive dogs should be sedated or tranquilized to facilitate the restraint necessary for smooth intravenous injection. Intravenous injection is the preferred route of administration because venipuncture is easily performed on most dogs by trained, experienced personnel. Injection outside the circulatory system is less reliable, is potentially painful, and almost invariably produces a slow onset of action.

Injectable drugs—such as magnesium sulfate, potassium chloride, and neuromuscular blocking agents (e.g., atracurium, curare, gallamine, pancuronium, succinylcholine, and vecuronium)—may be used (Bowen et al., 1970;

Hicks and Bailey, 1978); however, the dogs must be in a deep plane of anesthesia before drug administration (AVMA, 1993). Strychnine and nicotine are not suitable for euthanasia, because their stimulant properties might cause distress even in anesthetized animals.

Inhalation Methods

Overdose of a potent inhalant anesthetic (e.g., halothane and isoflurane) is satisfactory for performing euthanasia on dogs and is particularly appropriate for young dogs, in which venipuncture can be difficult. Anesthetic vapors tend to be irritating; therefore, the animals should be tranquilized first. If anesthetic vapors are used, a system for scavenging excess vapor is necessary to comply with federal guidelines on anesthetic-vapor pollution (CDC, 1977). Ether, unlike most contemporary inhalant anesthetics, is flammable and explosive; therefore, its use is not recommended.

Carbon monoxide and carbon dioxide both cause death by hypoxia. Carbon monoxide is impractical in most instances because of the risk to operators and the complexity of the equipment to administer it. Carbon dioxide has anesthetic properties and can be used for euthanasia (Carding, 1968; Leake and Waters, 1929); however, unless the chamber is well designed and used properly, dogs can become distressed before becoming unconscious. Hypoxia is not satisfactory for euthanatizing pups because young animals tolerate hypoxia better than older dogs and can survive for more than 30 minutes (Glass et al., 1944).

Physical Methods

Exsanguination is acceptable for euthanasia; however, the dog must be anesthetized because the decreasing blood flow causes anxiety and autonomic stimulation (Gregory and Wotton, 1984). Electrocution is considered a humane method of euthanasia, provided that sufficient current passes through the animal's brain to produce unconsciousness before or coincidentally with the onset of cardiac arrest. However, this method of euthanasia is not practical in most laboratories because of the danger to personnel (AVMA, 1993; Roberts, 1954; Warrington, 1974). Decapitation of pups is not recommended by the AVMA Panel on Euthanasia (1993).

Human Considerations

Euthanasia of dogs or any other animals can be stressful for the personnel performing the procedure. The degree of distress experienced by people observing or performing euthanasia depends on their backgrounds, personal philosophies, and ethical views on the use of animals in research (Arluke,

1988). People often transfer to the death of animals their unpleasant reactions to human death, and their responses to euthanasia can be magnified when strong bonds exist between them and the dogs being killed (e.g., strong bonds often develop between animal-care personnel and seriously ill canine models that require a great deal of care and rely totally on their human guardians). The stress experienced can be manifested as absenteeism, belligerence, careless and callous handling of animals, and high turnover rate. To be responsive to those concerns, institutional officials and supervisors should be aware of and sensitive to the issues and should provide opportunities for individual and group discussion and support and for educational programs that furnish factual information about euthanasia and teach stress-management and coping skills (NRC, 1991).

REFERENCES

Acha, P. N., and B. Szyfres. 1987. Rabies. Pp. 425-449 in Zoonoses and Communicable Diseases Common to Man and Animals, 2d ed. Scientific Pub. No. 503. Washington, D.C.: Pan American Health Organization.

Amyx, H. L. 1987. Control of animal pain and distress in antibody production and infectious disease studies. J. Am. Vet. Med. Assoc. 191:1287-1289.

Appel, M. J., ed. 1987. Virus Infections of Carnivores. Amsterdam: Elsevier Science Publishers. 500 pp.

Arluke, A. B. 1988. Sacrificial symbolism in animal experimentation. Object or Pet? Anthrozoös 2(2):98-117.

Auerbach, C. 1967. The chemical production of mutations. Science 158:1141-1147.

AVMA (American Veterinary Medical Association). 1993. 1993 Report of the AVMA Panel on Euthanasia. J. Am. Vet. Med. Assoc. 202:229-249.

Baker, J. A. 1970. Measles vaccine for protection of dogs against canine distemper. J. Am. Vet. Med. Assoc. 156:1743-1746.

Baker, J. A., D. S. Robson, L. E. Carmichael, J. H. Gillespie, and B. Hildreth. 1961. Control procedures for infectious diseases of dogs. Proc. Anim. Care Panel 11:234-244.

Barlough, J. E., ed. 1988. Manual of Small Animal Infectious Diseases. New York: Churchill Livingstone. 444 pp.

Booth, N. H. 1988a. Section 4: Drugs acting on the central nervous system. Pp. 153-405 in Veterinary Pharmacology and Therapeutics, 6th ed., N. H. Booth and L. E. McDonald, eds. Ames: Iowa State University Press.

Booth, N. H. 1988b. Intravenous and other parenteral anesthetics. Pp. 212-274 in Veterinary Pharmacology and Therapeutics, 6th ed., N. H. Booth and L. E. McDonald, eds. Ames: Iowa State University Press.

Bowen, J. M., D. M. Blackmon, and J. E. Haevner. 1970.   Effect of magnesium ions on neuromuscular transmission in the horse, steer, and dog. J. Am. Vet. Med. Assoc. 157:164-173.

Carding, A. H. 1968. Mass euthanasia of dogs with carbon monoxide and/or carbon dioxide; preliminary trials. J. Small Anim. Pract. 9:245-259.

Carmichael, L. E. 1979. Brucellosis (*Brucella canis*). Pp. 185-194 in CRC Handbook Series in Zoonoses, vol. 1, J. H. Steele, ed. Boca Raton, Fla.: CRC Press.

Carmichael, L. E. 1983. Immunization strategies in puppies—why failures? Compend. Contin. Educ. Practicing Vet. 5:1043-1051.

Carmichael, L. E. 1990. *Brucella canis.* Pp. 335-350 in Animal Brucellosis, K. Nielsen and J. R. Duncan, eds. Boca Raton, Fla.: CRC Press.

Carmichael, L. E., and C. F. Greene. 1990a. Canine herpesvirus infection. Pp. 252-258 in Infectious Diseases of the Dog and Cat, C. E. Greene, ed. Philadelphia: W. B. Saunders.

Carmichael, L. E., and C. E. Greene. 1990b. Canine brucellosis. Pp. 573-584 in Infectious Diseases of the Dog and Cat, C. E. Greene, ed. Philadelphia: W. B. Saunders.

CDC (Centers for Disease Control). 1977. Criteria for a Recommended Standard Occupational Exposure to Waste Anesthetic Gases and Vapors. HEW Pub. No. NIOSH 77-140. Washington, D.C.: U.S. Department of Health, Education, and Welfare. 194 pp. Available by interlibrary loan from the CDC Information Center, M/S C04, Atlanta, GA 30333.

Classen, D. C., R. S. Evans, S. L. Pestotnik, S. D. Horn, R. L. Menlove, and J. P. Burke. 1992. The timing of prophylactic administration of antibiotics and the risk of surgical-wound infection. N. Eng. J. Med. 326:281-286.

Denhart, J. W. 1992. Xylazine reversal with yohimbine. Pp. 194-197 in Current Veterinary Therapy. XI. Small Animal Practice, R. W. Kirk and J. D. Bonagura, eds. Philadelphia: W. B. Saunders.

Flecknell, P. A. 1987. Special techniques. Pp. 59-74 in Laboratory Animal Anaesthesia. An Introduction for Research Workers and Technicians. London: Academic Press.

Georgi, J. R., and M. E. Georgi. 1992. Canine Clinical Parasitology. Philadelphia: Lea & Febiger. 227 pp.

Glass, H. G., F. F. Snyder, and E. Webster. 1944. The rate of decline in resistance to anoxia of rabbits, dogs and guinea pigs from the onset of viability to adult life. Am. J. Physiol. 140:609-615.

Gleed, R. D. 1987. Tranquilizers and sedatives. Pp. 16-27 in Principles & Practice of Veterinary Anesthesia, C. E. Short, ed. Baltimore: Williams & Wilkins.

Glickman, L. T., R. B. Grieve, E. B. Breitschwerdt, M. Mika-Grieve, G. J. Patronek, L. M. Domanski, C. R. Root, and J. B. Malone. 1984. Serologic pattern of canine heartworm (*Dirofilaria immitis*) infection. Am. J. Vet. Res. 45:1178-1183.

Greene, C. E., ed. 1990. Infectious Diseases of the Dog and Cat. Philadelphia: W. B. Saunders. 971 pp.

Gregory, N. G., and S. B. Wotton. 1984. Time to loss of brain responsiveness following exsanguination in calves. Res. Vet. Sci. 37:141-143.

Hall, L. W., and K. W. Clarke. 1991. Veterinary Anaesthesia, 9th ed. London: Bailliere Tindall. 410 pp.

Hart, B. L. 1985. Behavioral indications for phenothiazine and benzodiazepine tranquilizers in dogs. J. Am. Vet. Med. Assoc. 186:1192-1194.

Harvey, R. C., and J. Walberg. 1987. Special considerations for anesthesia and analgesia in research animals. Pp. 380-392 in Principles & Practice of Veterinary Anesthesia, C. E. Short, ed. Baltimore: Williams & Wilkins.

Hicks, T., and E. M. Bailey, Jr. 1978. Succinylcholine chloride as a euthanatizing agent in dogs. Am. J. Vet. Res. 39:1195-1197.

Hoskins, J. D. 1990. Veterinary Pediatrics: Dogs and Cats from Birth to Six Months. Philadelphia: W. B. Saunders. 556 pp.

Hughes, H. C., and C. M. Lang. 1983. Control of pain in dogs and cats. Pp. 207-216 in Animal Pain: Perception and Alleviation, R. L. Kitchell and H. H. Erickson, eds. Bethesda, Md.: American Physiological Society.

Kaneko, J. J., ed. 1989. Clinical Biochemistry of Domestic Animals, 4th ed. San Diego: Academic Press. 932 pp.

Leake, C. D., and R. M. Waters. 1929. The anesthetic properties of carbon dioxide. Curr. Res. Anesth. Analg. 8:17-19.

Loeb, W. F., and F. W. Quimby, eds. 1989. The Clinical Chemistry of Laboratory Animals. New York: Pergamon Press. 519 pp.

Lumb, W. V., and E. W. Jones. 1984. Veterinary Anesthesia, 2d ed. Philadelphia: Lea & Febiger. 693 pp.

Marder, A. R. 1991. Psychotropic drugs and behavioral therapy. Vet. Clin. N. Am. 21(2):329-342.

Mirvish, S. S. 1968. The carcinogenic action and metabolism of urethan and N-hydroxyurethan. Adv. Cancer Res. 11:1-42.

Muir, W. W., III, and J. A. E. Hubbell. 1989. Handbook of Veterinary Anesthesia. St. Louis: C. V. Mosby. 340 pp.

National Association of State Public Health Veterinarians. 1993. Compendium of animal rabies control, 1993. J. Am. Vet. Med. Assoc. 202:199-204.

NRC (National Research Council), Institute of Laboratory Animal Resources, Committee on Care and Use of Laboratory Animals. 1985. Guide for the Care and Use of Laboratory Animals. NIH Pub. No. 86-23. Washington, D.C.: U.S. Department of Heath and Human Services. 83 pp.

NRC (National Research Council), Institute of Laboratory Animal Resources, Committee on Educational Programs in Laboratory Animal Science. 1991. Euthanasia. Pp. 67-74 in Education and Training in the Care and Use of Laboratory Animals: A Guide for Developing Institutional Programs. Washington, D.C.: National Academy Press.

NRC (National Research Council), Institute of Laboratory Animal Resources, Committee on Pain and Distress in Laboratory Animals. 1992. Recognition and Alleviation of Pain and Distress in Laboratory Animals. Washington, D.C.: National Academy Press. 137 pp.

Popilskis, S., D. Kohn, J. A. Sanchez, and P. Gorman. 1991. Epidural *vs.* intramuscular oxymorphone analgesia after thoracotomy in dogs. Vet. Surg. 20:462-467.

Roberts, T. D. M. 1954. Cortical activity in electrocuted dogs. Vet. Rec. 66:561-566.

Rubal, B. J., and C. Buchanan. 1986. Supplemental chloralose anesthesia in morphine premedicated dogs. Lab. Anim. Sci. 36:59-64.

Scott, J. P. 1970. Critical periods for the development of social behaviour in dogs. Pp. 21-32 in The Post-Natal Development of Phenotype, S. Kazda and V. H. Denenberg, eds. Prague: Academia.

Sheffy, B. E., J. A. Baker, and J. H. Gillespie. 1961. A disease-free colony of dogs. Proc. Anim. Care Panel 11:208-214.

Short, C. E., ed. 1987. Principles & Practice of Veterinary Anesthesia. Baltimore: Williams & Wilkins. 669 pp.

Skarda, R. T. 1987. Local and regional analgesia. Pp. 91-133 in Principles & Practice of Veterinary Anesthesia, C. E. Short, ed. Baltimore: Williams & Wilkins.

Slattum, M. M., L. Maggio-Price, R. F. DiGiacomo, and R. G. Russell. 1991. Infusion-related sepsis in dogs undergoing acute cardiopulmonary surgery. Lab. Anim. Sci. 41:146-150.

Soma, L. R., ed. 1971. Textbook of Veterinary Anesthesia. Baltimore: Williams & Wilkins. 621 pp.

Soma, L. R. 1987. Assessment of animal pain in experimental animals. Lab. Anim. Sci. 37(Special Issue):71-74.

Soma, L. R., and D. R. Shields. 1964. Neuroleptanalgesia produced by fentanyl and droperidol. J. Am. Vet. Med. Assoc.145:897-902.

Swango, L. J. 1983. Canine Immunization. Pp. 1123-1127 in Current Veterinary Therapy. VIII. Small Animal Practice, R. W. Kirk, ed. Philadelphia: W. B. Saunders.

Thomas, E. D., and J. W. Ferrebee. 1961. Disease-free dogs for medical research. Proc. Anim. Care Panel 11:230-233.

Voith, V. L. 1984. Possible pharmacological approaches to treating behavioural problems in animals. Pp. 227-234 in Nutrition and Behaviour in Dogs and Cats, R. S. Anderson, ed. Oxford: Pergamon Press.

Wallach, M. B., K. E. Peterson, and R. K. Richards. 1981. Electrophysiologic studies of a combination of secobarbital and dibucaine for euthanasia of dogs. Am. J. Vet. Res. 42:850-853.

Warrington, R. 1974. Electrical stunning, a review of the literature. Vet. Bull. 44:617-628.

Willis, M. B. 1989. Genetics of the Dog. London: H. F. & G Witherby. 417 pp.

6

Special Considerations

PROTOCOL REVIEW

One of the many important responsibilities of an institutional animal care and use committee (IACUC) is to review the protocols of research projects in which dogs will be used (9 CFR 2.31; PHS, 1986). The protocol-review mechanism is designed to ensure that investigators consider the care and use of their animals and that protocols comply with federal, state, and institutional regulations and policies. In addition, the review mechanism enables an IACUC to become an important institutional resource, assisting investigators in all matters involving the use of animals. Although the discussion below is directed to the use of dogs in research, the review requirements apply to all vertebrate species.

Each research protocol must completely (but concisely) delineate the proposed study, including a description of each of the following:

- the purpose of the study;
- the rationale for selecting dogs as the research subjects;
- the breed, age, and sex of the dogs to be used;
- the numbers of dogs in various groups of the protocol and the total number to be used;
- experimental methods and manipulations;
- experimental manipulations that will be performed repeatedly on an individual dog;

- preprocedural and postprocedural care and medications;
- procedures that will be used to minimize discomfort, pain, and distress, including, where appropriate, the use of anesthetics, analgesics, tranquilizers, and comfortable restraining devices;
- the euthanasia method, including the reasons why it was selected and whether it is consistent with the recommendations of the American Veterinary Medical Association Panel on Euthanasia (AVMA, 1993, et seq.);
- the process undertaken to ensure that there are no appropriate in vitro alternatives, that there are no alternative methods that would decrease the number of animals to be used, and that the protocol does not unnecessarily duplicate previous work; and
- the qualifications of the investigators who will perform the procedures outlined.

One approach used by IACUCs is to have a scientifically knowledgeable member thoroughly review the protocol. The reviewer contacts the investigator directly to clarify issues in question. Later, at an IACUC meeting, the reviewer presents and discusses the protocol and relates discussions with the investigator. Changes or clarifications in the protocol that have resulted from the reviewer's discussions with the investigator are submitted to the IACUC in writing. After presentation of the protocol, the reviewer recommends a course of action, which is then voted on by the IACUC. Another kind of protocol review (which is especially effective in small institutions with few grants) is initial review by the entire IACUC; results are generally available to the investigator within a short period.

Several outcomes of protocol review are possible: approval, approval contingent on receipt of additional information (to respond to minor problems with the protocol), deferral and rereview after receipt of additional information (to respond to major problems with the protocol), and withholding of approval. If approval of a protocol is withheld, an investigator should be accorded due process and be given the opportunity to rebut the IACUC's critique in writing, to appear in person at an IACUC meeting to present his or her viewpoint, or both. It is also important that provision be made for expedited review, in which a decision is reached within 24-48 hours. Expedited reviews should be used only for emergency or extenuating circumstances. When a protocol is submitted for expedited review, each member of the IACUC must have an opportunity to review it and may call for a full committee review before approval is given and before animal work begins (McCarthy and Miller, 1990).

The question of protocol review for scientific merit has been handled in a variety of ways by IACUCs. Many protocols are subjected to extensive, external scientific review as part of the funding process; in such instances, the IACUC can be relatively assured of appropriate scientific review. In the case of studies that will not undergo outside review for scientific merit, many IACUCs require signoff by the investigators, department chairmen, or internal review committees; this makes the signer responsible for providing assurance that the proposed studies have been designed and will be performed "with due consideration of their relevance to human or animal health, the advancement of knowledge, or the good of society" (NRC, 1985, p. 82; PHS, 1986, p. 27). Occasionally, IACUC members and investigators differ as to the relevance of proposed studies to human and animal health and the advancement of knowledge. Each institution should develop guidelines for dealing with this potential conflict.

RESTRAINT

Some form of restraint is generally necessary to control a dog during a procedure (see guidelines in NRC, 1985, p. 9). The method used should provide the least restraint required to allow the specific procedure to be performed properly, should protect both the dogs and personnel from harm, and should avoid causing distress, physical harm, or unnecessary discomfort. In handling and restraining dogs, it is helpful to understand species-typical behavior patterns and communication systems.

A small or medium-size dog can be picked up by placing one hand under the chest and abdomen while restraining the head with a leash. Lifting a large dog might require two people. It is important to remember that males are sensitive to touch near their genitalia. Minor procedures, such as taking a rectal temperature or administering a subcutaneous injection, can usually be accomplished by one person using minimal restraint. During venipuncture, sufficient restraint should be used to avoid repeated needle insertions and to prevent the development of painful hematomas. Kesel and Neil (1990) detail methods for handling and restraining animals.

If dogs are to be restrained frequently or for long periods or if the restraint method used is especially rigorous, it might be necessary to train them to tolerate the restraint. Training sessions should use positive-reinforcement techniques; negative-reinforcement techniques are not desirable. Physical abuse (9 CFR 2.38f2i) and food or water deprivation (9 CFR 2.38f2ii) must not be used to train, work, or handle dogs, although food and water may be withheld for short periods when specified in an IACUC-approved protocol (9 CFR 2.38f2ii).

SPECIAL CARE FOR ANIMAL MODELS

The remainder of this chapter deals with some common uses of laboratory dogs in which aspects of care vary from the general guidelines provided in previous chapters. It is not intended to present an exhaustive list of canine models that require special housing and husbandry, but rather to provide the reader with different types of canine models that can serve as examples of how housing and husbandry can be modified to achieve animal well-being. The suggestions offered here are not to be construed as the only ones possible. The committee recognizes that not every research procedure and circumstance can be anticipated, and it assumes that sound professional judgment, good veterinary practices, and adherence to the spirit of this guide will prevail in unusual situations.

The final subsection of this chapter introduces the reader to the technique of somatic cell gene therapy. Many disorders of dogs, like those of humans, are caused by single-gene mutations. Scientists are working to develop techniques to cure these disorders permanently by replacing mutant genes with normal ones. For many reasons (see Chapter 2), the dog is an ideal model for evaluating the safety and efficacy of gene therapy.

Aging

Clinical Features

Life expectancy and disease incidences vary among breeds of dogs; therefore, it is not possible to state a specific age at which dogs become old. Common laboratory dogs, such as beagles, begin some aging changes when they are 8-10 years old. Such physical features as graying of the haircoat, especially around the face, are often apparent as aging begins.

As dogs age, they tend to become less active and to exhibit such signs of mental deterioration as poor recognition of caretakers, excessive sleeping, and changes in personality. Senile plaques, similar to those found in humans with senile dementias, have been reported in the brains of old dogs (Wiśniewski et al., 1970). Various forms of arthritis, spondylosis, and degenerative joint disease are common and contribute to problems in mobility and to the apparent diminution of mental alertness. Older dogs might decrease their daily food intake, become slow eaters, or become irregular in their eating habits. Dental problems—including periodontal disease, tooth abscesses, and oral-nasal fistulas—increase; the importance of these problems is probably underestimated (Tholen and Hoyt, 1983). Dogs more than 6 years old develop lenticular sclerosis, which results in a bluish appearance within the pupil. Visual acuity decreases with age and is often associated with cataracts, secondary glaucoma, and other diseases (Fischer, 1989). There is also apparent hearing loss.

Atrophy of the thyroid gland and an increased number of thyroid tumors have been reported, and signs of hypothyroidism are common (Haley et al., 1989; Milne and Hayes, 1981). Thyroid atrophy and the propensity of older dogs to develop hypothermia might be related (B. A. Muggenburg,

Inhalation Toxicology Research Institute, Lovelace Biomedical and Environmental Research Institute, Albuquerque, N.M., unpublished). A decreased response to antigens and changes in lymphocyte function might indicate that the older dog is less able to resist infectious diseases (Bice and Muggenburg, 1985). Some changes in common blood-cell measures and serum chemistry become important when these are used for diagnosis (Lowseth et al., 1990a). The incidence of neoplasia increases strikingly (MacVean et al., 1978); for example, lung tumors, nearly unknown in young dogs, can reach an incidence as high as 10 percent in dogs over 10 years old (Ogilive et al., 1989). Pulmonary function decreases with age because of reduced lung volumes and decreased elasticity (Mauderly and Hahn, 1982). Chronic renal diseases often occur and require frequent monitoring. Chronic heart disease is also fairly common, and clinical signs can appear suddenly in old dogs.

Husbandry and Veterinary Care

Housing and environment. Accommodation should be made for dogs that have problems moving comfortably on floor grates or through guillotine-like doors in kennel buildings. Because of their decreased mobility and impaired thermoregulatory function, aging dogs with access to outdoor areas should be checked frequently to be certain that they are able to get inside to escape the cold or heat. Automatic watering devices might become difficult to use; for some old dogs, it might be necessary to switch to water pans placed on the floor.

Nutrition. Differentiation between age-related and disease-caused changes in eating habits might be difficult. It is important that animal-care personnel become familiar with and closely monitor daily eating habits of older dogs. Frequent checking and recording of body weights can help in assessing whether food intake is adequate. Changes in diet are sometimes dictated by the clinical diagnosis of disease (e.g., a low-protein diet for chronic, progressive renal disease and a low-sodium diet for chronic heart failure).

Physical characteristics of food can affect dental hygiene. Soft and wet food fed over many years can contribute to dental disease. Feeding dry dog food and providing hard objects for chewing can be helpful in the long-term management of dental problems. Routine dental care, including the removal of calculus and polishing, is essential.

Veterinary care. The extent of chronic disease problems in older dogs requires more intensive veterinary care, extensive diagnostic investigations, and good nursing. Dosages of some medications might have to be reduced, because drugs are commonly metabolized more slowly in old than in young adult dogs. Such drugs as digoxin should be monitored by measuring blood concentrations to decrease the risk of overdosing (De Rick et al., 1978). A useful reference on geriatric veterinary medicine is *Geriatrics and Gerontology* (Goldston, 1989).

Reproduction

Bitches. Andersen and Simpson (1973) have described reproductive senescence in beagle bitches. Intact bitches exhibit irregular estrous cycles, accompanied by decreased fertility, and prolonged periods of anestrus. The mortality rate is higher among puppies born to older bitches than among puppies born to bitches less than 3 years old.

The most common pathologic condition of the uterus of aged bitches is pyometra (Andersen and Simpson, 1973; Järvinen, 1981; Whitney, 1967). Vaginal fibromuscular polyps are also common (Andersen and Simpson, 1973). The age-specific incidence of mammary gland neoplasms in intact beagle bitches continues to increase throughout life (Taylor et al., 1976).

Dogs. Aging dogs have testicular atrophy and often develop prostatic hypertrophy and hyperplasia and have episodes of prostatitis (Lowseth et al., 1990b). There are also metaplastic changes in the bladder (Lage et al., 1989).

Cardiovascular Diseases

Congenital Heart Defects

Clinical Features

Dogs with hereditary cardiovascular malformations have been used to investigate the role of genetic and embryologic factors in the cause and pathogenesis of congenital heart defects, including hereditary patent ductus arteriosus, conotruncal defects (e.g., ventricular septal defect, tetralogy of Fallot, and persistent truncus arteriosus), discrete subaortic stenosis, and pulmonary valve dysplasia. Congenital heart defects in dogs have been summarized by Buchanan (1992) and Eyster (1992). Table 6.1 describes and lists the clinical signs of selected heart defects. Each of those defects is transmitted as a lesion-specific genetic defect in one or more breeds. A model for each defect has been developed at the University of Pennsylvania School of Veterinary Medicine by selective breeding of affected dogs (Patterson, 1968), as follows: patent ductus arteriosus, toy and miniature poodles (Ackerman et al., 1978; De Reeder et al., 1988; Gittenberger-de Groot et al., 1985; Knight et al., 1973; Patterson et al., 1971); conotruncal defects, keeshonden (Patterson et al., 1974, 1993; Van Mierop et al., 1977); discrete subaortic

TABLE 6.1   Selected Congenital Cardiac Defects in Dogs

Defect	Description	Clinical Signs
Patent ductus arteriosus	Failure of ductus arteriosus to close after birth. If pulmonary vascular resistance is low, blood flows through ductus from left to right. Pulmonary hypertension and left ventricular hypertrophy result unless ductus opening is small. If ductus is large and pulmonary vascular resistance is high, pulmonary arterial pressure can exceed aortic pressure, and blood will flow from right to left, sending venous blood into ascending aorta.	Vary with size of duct and pulmonary vascular resistance from subclinical to heart failure. Early signs include poor growth, coughing, and dyspnea. Aneurysm can occur at site of ductus arteriosus. Polycythemia occurs in cyanotic dogs with a large patent ductus arteriosus (PDA), pulmonary hypertension, and right to left blood flow through the PDA.
Conotruncal defects		
Ventricular septal defect	Failure to complete formation of the conotruncal septum results in ventricular septal defects (VSDs) of varied size, involving the lower and middle portions of the crista supraventricularis (Type I, subarterial VSD). Pups with large VSDs usually die from pulmonary edema in the neonatal period. Smaller VSDs are compatible with long life unless complicated by pulmonary hypertension and congestive heart failure.	Vary with size of defect from subclinical to signs of respiratory and right-side heart failure, including cyanosis, dyspnea, weakness, and anorexia.
Tetralogy of Fallot	Consists of pulmonic stenosis (valvular, infundibular, or both), conal ventricular septal defects, dextroposition of aorta with overriding of ventricular septum, and right ventricular hypertrophy. Some dogs have pulmonary valve atresia (pseudo-truncus arteriosus).	Depend on severity of pulmonic stenosis and ventricular septal defect. Can include decreased body size, fatigue, cyanosis, and secondary polycythemia.
Persistent truncus arteriosus	Severe but rare anomaly. Complete failure of septation of conus and truncus regions, producing large conal ventricular septal defect and single arterial outlet vessel.	Cyanosis and dyspnea. Dogs rarely survive neonatal period.

TABLE 6.1 *Continued*

Defect	Description	Clinical Signs
Discrete subaortic stenosis	Narrowing of left ventricular outflow tract, most commonly by fibrous ring just below aortic semilunar valves, with concomitant obstruction of blood flow, left ventricular hypertrophy, and increased left ventricular pressure.	Vary with degree of stenosis from asymptomatic to poor growth, exercise intolerance, syncope, ventricular arrhythmias, pulmonary edema, and sudden death.
Pulmonary valve dysplasia	Varies from mild thickening of leaflets surrounding narrowed pulmonary orifice to complete fusion of leaflets and doming of valve. Interferes with emptying of right ventricle.	Vary from asymptomatic to dyspnea, fatigability, and right-side heart failure.

stenosis, Newfoundlands (Patterson, 1984; Pyle et al., 1976); and pulmonary valve dysplasia, beagles (Patterson, 1984; Patterson et al., 1981). Conotruncal defects in the keeshond breed are determined by the effect of a single major gene defect (Patterson et al., 1993). Subaortic stenosis in Newfoundlands also appears to be monogenic with variable expression (Patterson, 1984). Patent ductus arteriosus and pulmonary valve dysplasia are inherited in a non-Mendelian pattern.

Husbandry and Veterinary Care

Animals with cardiac defects often require exercise restriction to avoid cyanosis and congestive heart failure. The need for restriction must be decided for each dog on the basis of cardiac status. If the clinical manifestations of severe defects (e.g., respiratory distress, severe cyanosis, and congestive heart failure) cannot be relieved with appropriate surgical methods or cardiovascular drugs (e.g., cardiac glycosides and diuretics), the dog should be humanely killed (see Chapter 5).

Reproduction

Only dogs with mild to moderate cardiac defects or those in which the defects have been surgically corrected should be selected for breeding. Severely affected dogs do not survive to breeding age, or they develop clinical manifestations that preclude their use for reproduction (e.g., marked cyanosis and congestive heart failure). Methods of modern clinical cardiology—including auscultation, radiography, echocardiography, cardiac catheterization, and angiocardiography—are necessary for accurate diagnosis and evaluation of the severity of defects in candidates for breeding. Therefore, appropriate facilities and equipment and personnel qualified to use such equipment must be available before a breeding colony is established. Once it is established, the health status of breeding stock and their offspring must be carefully monitored.

Induced Heart Defects

Clinical Features

Many animal models of cardiac disease are surgically induced in physiologically normal animals. Aims of the research protocol and humane considerations must often be carefully balanced to ensure that the maximal amount of information is derived from each animal.

Surgically induced models can be broadly divided into models of volume or pressure overload produced by creating valvular or interchamber defects, models of ischemic injury, and models of arrhythmia (Gardner and Johnson, 1988). Long-term management of these models can be difficult because they are frequently on the verge of physiologic decompensation and at risk of sudden death. Table 6.2 lists the signs of cardiac failure.

TABLE 6.2   Clinical Signs of Heart Failure in Dogs

Type of Heart Failure	Clinical Signs
Left-side	Exercise tolerance decreases. Inappropriate dyspnea follows exercise. Pulmonary venous pressure increases, initially causing pulmonary and bronchial congestion and reflexogenic bronchoconstriction. Repetitive coughing follows exercise. Orthopnea, with a reluctance to lie down; restlessness at night; and paroxysmal dyspnea are common. In severe failure, pulmonary edema, severe dyspnea at rest, and rales on auscultation become evident.
Right-side	Systemic venous congestion occurs with engorgement of jugular veins. Liver and spleen are enlarged and often palpable. Fluid retention is usually first manifested as ascites; subcutantous edema, hydrothorax, or hydropericardium can follow. Disturbances of gastrointestinal function, with diarrhea, can occur.
Generalized	Signs of both left- and right-side failure occur.

Husbandry and Veterinary Care

The management of chronic dog models of induced heart failure is most successful if the approach used is interdisciplinary, involving cardiologists, surgeons, and veterinary-care staff. Goals of long-term management include identifying potential complications, selecting therapeutic regimens, and developing long-term monitoring protocols. The following general guidelines should be tailored to the type of disorder induced, the dogs' well-being, and the goals of the research protocol.

Postoperative care. Postoperative care depends on the type of heart disease induced. Medical management should continue after successful recovery from surgery because a specific surgical protocol does not always produce a physiologically consistent model. Some dogs achieve a stable, compensated postoperative condition; others undergoing the same procedure develop signs of acute heart failure immediately after surgery.

Careful monitoring on the days after surgery is critical. Meticulous physical examinations should be performed on physiologically stable dogs at least once a day until they have recovered from surgery. Physiologically unstable dogs should be examined more often. Vital signs should be monitored, and particular attention should be given to physical findings related to the cardiovascular system. Mucous membrane color, capillary-refill time, and temperature of extremities can be abnormal if peripheral perfusion is seriously impaired. The pulse quality of the femoral artery can be used to assess systemic perfusion. Auscultation should be used to detect abnormal cardiac sounds, and electrocardiography should be performed to diagnose arrhythmias. Assessment of respiratory rate and depth should be combined with careful auscultation of all lung fields to detect early signs of pulmonary complications. Echocardiography, if available, can be used to evaluate cardiac function and contractility.

Good nursing care is important. Special diets, such as canned dog food or dry food mixed with chicken broth, can be offered to encourage food intake. Ideally, dogs should be housed in a dedicated recovery room and returned to the regular housing area only when they are physiologically stable and have recovered fully from surgery. Decreased exercise tolerance secondary to diminished cardiac reserve might affect the extent of activity that a dog can withstand.

Complications. Potential complications associated with surgical and catheterization procedures should be anticipated, including infection of the operative site, bacteremia, and endocarditis. Dogs at high risk for complications are the ones that undergo serial catheterization procedures and those with bioimplants, such as prosthetic valves and pacemakers (Dougherty,

1986). Baseline monitoring should include scheduled physical examinations and complete blood counts (CBCs). A blood culture should be submitted to the laboratory for any animal with a persistent fever or an intermittently increased temperature. If infection is suspected, a broad-spectrum antibiotic, such as one of the cephalosporins, should be administered pending receipt of culture and sensitivity results.

Banding of the great vessels with various materials is a standard procedure for producing volume- and pressure-overload models of ventricular hypertrophy, coarctation of the aorta, and obstruction of right ventricular outflow. Vessel erosion caused by the material used (Gardner and Johnson, 1988) and hemorrhage secondary to banding procedures are common complications that should be included in the differential diagnosis of any banded animal that suffers an acute onset of lethargy, paleness of the mucous membranes, or dyspnea. Those are also clinical signs of heart failure, so it is important to perform auscultation of the chest and suitable diagnostic tests, such as radiography or thoracentesis, to make an accurate diagnosis. A dog that is hemorrhaging should be euthanatized.

Surgical procedures used to induce cardiac disease invariably cause disruption of the endothelium and put the dogs at risk for thrombosis and embolism. Dogs undergoing cardiac catheterization or surgery of the cardiac valves are at greatest risk. Clinical signs reflect the organs involved.

Long-term monitoring. In a study of extended duration, assessment of each dog's general health and cardiovascular system should be continuous. The type and frequency of examinations will depend on whether the model is physiologically stable or unstable. For example, a dog with induced mitral regurgitation, which is defined as a 50 percent reduction in forward stroke volume and a pulmonary capillary wedge pressure of 20 mm Hg, can develop life-threatening pulmonary edema (Nakano et al., 1991; Swindle et al., 1991). Frequent monitoring and auscultation are required to detect early signs of respiratory compromise so that the dog will not die before therapy can be initiated or the dog can be studied. Similarly, a dog with induced right ventricular pressure overload requires frequent monitoring because decreased coronary blood flow can lead to acute right-side heart failure (Fixler et al., 1973; Vlahakes et al., 1981). Conversely, a stable model of left ventricular hypertrophy can be produced in 8-week-old pups by aortic banding, which causes a systolic pressure gradient of 15-20 mm Hg (O'Kane et al., 1973). Dogs with induced tricuspid valve insufficiency can tolerate increased venous pressure and a slight reduction in cardiac output for years, although some develop ascites and reduced serum albumin (Arbulu et al., 1975). These models require less frequent monitoring.

Equipment. Follow-up care and monitoring require appropriate equipment and laboratory support for obtaining CBCs, blood cultures, serum chemistry profiles, and blood-gas analyses. Electrocardiography and echocardiography should be available for assessing cardiac rhythm and function, respectively. Echocardiography is also a useful noninvasive method for monitoring changes in cardiac wall thickness, cardiac motion, and chamber size as cardiac disease progresses. A cardiac catheterization laboratory should be available for performing hemodynamic and angiographic studies.

Pharmacologic therapy. Pharmacologic management of dogs that develop complications or clinical signs of heart failure must be coordinated between the veterinary unit and the investigator to prevent the administration of medications that could compromise the scientific aims of the study. Diuretics can be used to treat pulmonary edema and reduce plasma volume, but their effects on serum electrolytes and the reduction of venous return and cardiac output should be considered. Vasodilators, calcium antagonists, β-blocking drugs, and positive inotropic agents should be available for managing acute clinical events; however, long-term use of these drugs is usually contraindicated because of their effects on the disease process being studied (Bonagura, 1986; Swindle et al., 1991).

Hypertension

Clinical Features

To provide proper care for hypertensive dogs and to avoid inappropriate treatment that can be detrimental to the dogs and compromise the study, it is necessary to have a full understanding of the pathophysiology of hypertension and of the specific method that is used to induce it. Generally, hypertension in dogs is induced by constricting the renal artery. The resulting reduction in renal perfusion causes systemic arterial pressure—and renal arterial pressure distal to the constriction—to rise enough to maintain renal function. A discussion of the relationship between renal function and the long-term control of blood pressure can be found in any standard physiology textbook (e.g., Guyton, 1991).

Two methods are most commonly used to induce renal vascular hypertension: partial constriction of one renal artery (the 2-kidney, 1-clip method) and unilateral nephrectomy and partial constriction of the remaining renal artery (the 1-kidney, 1-clip method). Both those methods produce what is called Goldblatt hypertension, but the mechanisms responsible for the hypertension are different. The 2-kidney, 1-clip model depends more heavily on the renin-angiotensin system than the 1-kidney, 1-clip model and responds to acute treatment with angiotensin-converting enzyme (ACE) inhibitors, which block the conversion of angiotensin I to angiotensin II. The

1-kidney, 1-clip model requires chronic treatment with ACE inhibitors to lower blood pressure. The reason for that difference is described in detail by Guyton (1991).

The greatest success in producing hypertension while reducing the incidence of malignant hypertension and renal failure is achieved by reducing renal arterial flow by exactly 50 percent. Renal blood flow is usually measured when the arterial clamp (Goldblatt clamp) is adjusted during surgery; this obviates later surgery to readjust the degree of constriction. Methods have been developed for measuring renal blood flow chronically and adjusting the renal artery clamp (Ferrario et al., 1971), and more recently a technique has been described for producing hypertension reliably by gradually constricting the renal artery with constrictors fabricated of ameroid, a hydroscopic material made of compressed casein cured in formalin (Ben et al., 1984; Brooks and Fredrickson, 1992).

Other methods that have been used for inducing hypertension include a 2-kidney, 2-clip model in which Goldblatt clamps or ameroid constrictors are applied to both renal arteries; wrapping of one or both kidneys with silk or cellophane; a combination of unilateral nephrectomy and wrapping of one kidney; and placing sutures in a figure 8 configuration on the surface of one or both kidneys (the Grollman model). The creation of hypertension with deoxycorticosterone acetate (DOCA) and common salt has not been as successful in dogs as it has in rats, because dogs are reluctant to eat a high-salt diet or drink a saline solution. However, moderate hypertension in dogs can be achieved with DOCA administration alone. A colony of spontaneously hypertensive dogs has been described (Bovée et al., 1986).

Husbandry and Veterinary Care

Proper care of hypertensive dogs involves the following:

- careful design and establishment of the hypertensive model to produce stable hypertension;
- routine evaluation of renal function;
- regular and frequent monitoring of blood pressure;
- regular monitoring of the retinas;
- appropriate treatment with antihypertensives when required; and
- careful husbandry.

Evaluation of renal function. Routine evaluation of renal function is essential because renal failure is a common complication in dogs with experimental hypertension. Renal failure can be caused by too much constriction of the renal artery, a rapid increase in both systolic and diastolic pressures (malignant hypertension), or the inappropriate use of antihypertensives.

Evaluation of renal function is especially important with use of the 1-kidney, 1-clip and 2-kidney, 2-clip models (which cause the most severe hypertension) and during antihypertensive therapy. In hypertensive dogs, renal function is compromised to such an extent that blood pressure must be raised to maintain sodium balance. If antihypertensive therapy lowers blood pressure too much, acute renal failure will ensue.

The most reliable and easily measured indicators of renal function are serum creatinine concentration and blood urea nitrogen (BUN). Although they depend somewhat on the type of assay, normal serum creatinine for the dog ranges between 0.4 and 1.3 mg/dL and BUN between 10 and 25 mg/dL. Serum creatinine and BUN should be determined in each dog before hypertension is induced to avoid using dogs with already-compromised renal function. In Goldblatt hypertensive models, serum creatinine should be determined daily for the first 5 days after surgery and twice a week thereafter. If ameroid constrictors are used, daily evaluations should continue through the second week after surgery because it takes 4-5 days for ameroid constrictors to reach maximal constriction. In models in which hypertension is not as severe, such as the 2-kidney, 1-clip and 1-kidney, 1-wrapped hypertensive models, renal function is less likely to be impaired, and serum creatinine concentration and BUN might not be increased, but they should be evaluated at least once during the 10-day postoperative period.

If renal-function tests show signs of renal failure, corrective action should be taken. Too-severe constriction of the renal artery can be corrected surgically, or the study can be terminated by euthanatizing the animal. Renal failure caused by lowering blood pressure to below the renal autoregulatory range should be corrected by reducing the dose of the antihypertensive drug to a point that allows blood pressure to remain high enough to maintain renal function. Malignant hypertension can be treated with antihypertensives and reduced salt intake (Ross, 1989; see below).

Measurement of arterial blood pressure. Blood pressure should be determined routinely after surgery. It can be done with indirect methods, such as placing a pressure cuff at the base of the tail (Petersen et al., 1988) or above the hock, or with direct methods, such as chronic implantation of arterial catheters or acute femoral arterial catheterization. To avoid complications associated with exteriorized catheters, some investigators now use methods that do not require exteriorized components, such as a Vascular Access Port (Access Technologies, Skokie, Ill.) (Mann et al., 1987), or chronic instrumentation, such as constriction of the carotid loop (Brooks et al., 1991). In addition, improved telemetric monitoring (Lange et al., 1991) has the potential to allow continuous monitoring of blood pressure over a number of days or weeks.

It is important to establish a baseline blood pressure before inducing hypertension. Measuring blood pressure several times permits the dog to become accustomed to the monitoring technique and thereby avoids increases in blood pressure caused by stress. Some investigators measure blood pressure indirectly (e.g., with the tailcuff method) before surgery and use more direct methods later. That is done in recognition that indirect methods can lead to a deviation of up to 10 mm Hg from true arterial pressure. Normal systolic blood pressure ranges from 112 to 142 mm Hg; normal diastolic pressure from 56 to 110 mg Hg. Measurements greater than 160/95 indicate hypertension.

Treatment for hypertension. When induced correctly, surgically created hypertension is sustained and has few complications. If necessary, hypertensive dogs can be maintained on special diets (see below) and given diuretics or other antihypertensive drugs when needed. Some drugs readily available for treatment of hypertensive dogs are listed in Table 6.3.

Malignant hypertension must be diagnosed quickly and treated aggressively. The most striking clinical sign of malignant hypertension can be blindness caused by retinal detachment, which is usually preceded by retinal hemorrhage, dilation of retinal vessels, and subretinal edema. The dogs do not appear to be in pain but often bump into walls and might become disoriented or sit quietly in their pens or cages. Diagnosis can easily be confirmed with an ophthalmologic examination. If blood pressure can be controlled and retinal disinsertion (detachment from the ora ciliaris retinae)

TABLE 6.3 Drugs Available for the Oral Treatment of Hypertension in Dogs[a]

Generic Name	Dosage, mg/kg	Frequency of Administration	Class
Chlorothiazide	20-40	Every 12 hr or daily	Diuretic
Hydrochlorothiazide	2-4	Every 12 hr or daily	Diuretic
Furosemide[b]	2-4	Every 8-12 hr	Diuretic
Propranolol	0.25-0.5	Every 8 hr	β-Adrenergic antagonist
Hydralazine	1-3	Every 12 hr	Vasodilator
Prazosin	0.25-2	Every 8 hr	Vasodilator
Verapamil[c]	1-2	Every 8 hr	Vasodilator; calcium-channel blocker
Captopril	0.5-1	Every 8-12 hr	Angiotensin-converting enzyme inhibitor

[a]Adapted from Ross, 1989; printed with permission of the author and W. B. Saunders, Philadelphia, Pennsylvania.

[b]Can also be given intramuscularly or intravenously at 2-4 mg/kg.

[c]Can also be given intravenously at 0.05-0.15 mg/kg.

does not occur, some vision might be restored in 2-3 weeks. Malignant hypertension often responds well to treatment with ACE inhibitors. Diuretics can also be administered if care is taken to avoid a precipitous drop in renal blood flow. Vasodilators can be used with caution. If the cause of the malignant hypertension is overconstriction of the renal artery, ACE inhibitors can be used to stabilize the dog while the stricture is surgically corrected.

Husbandry. Routine care of hypertensive dogs must include a consideration of diet because both salt intake and protein intake will affect blood pressure and renal function. A high salt intake will exacerbate hypertension, and a high protein intake might accelerate the loss of renal function. To avoid unintended changes in diet that could compromise their dogs and studies, investigators, veterinarians, and others caring for hypertensive dogs should establish dietary requirements before beginning studies.

For dogs with hypertension and renal failure, the diet should contain 0.1-0.3 percent sodium on a dry-weight basis or 10-40 mg/kg per day (5-20 mg/lb per day) (Ross, 1989). Low-protein diets (less than 15 percent) that are also low in sodium (e.g., K/D, Hill's Pet Products, Inc., Topeka, Kansas) are available and should be fed in adequate amounts—generally 1 can or 2 cups of dry food for each 10 kg (20 lb) of presurgical body weight. The protein content of some commercially available diets might be too low to maintain ideal body weight, but diets that combine a higher protein content with a lower sodium content are available (e.g., R/D, Hill's Pet Products, Inc., Topeka, Kansas). As in any dog model, following the body weight of an animal regularly is a good way to monitor the animal's overall health.

There is usually no reason to restrict primary enclosure size for hypertensive dogs. Whether they should be exempted from an exercise program depends on their postoperative course. If the hypertensive condition stabilizes and there are no complications, exemption from exercise should not be necessary. Blood pressure is known to increase in stressful conditions; therefore, it is important that such conditions be avoided (e.g., dogs that are housed or exercised in pairs or groups should be monitored to ensure that they are compatible).

Ehlers-Danlos Syndrome

Clinical Features

Ehlers-Danlos syndrome type 1 is an autosomal dominant condition of humans for which there are analogues in dogs and other mammals (Hegreberg et al., 1969, 1970). The disease is caused by a defect in metabolism of dermal collagen that results in a skin tensile strength less than 10 percent of normal. Fibrous tissue and bone are subclinically affected in some cases (Minor et al., 1987). Multiple lacerations are often observed. The hyperextensible skin can cause superior entropion, inferior ectropion, or both.

Husbandry and Veterinary Care

The extreme fragility of the skin must be considered in managing dogs with this syndrome. Affected dogs should be housed singly in smooth-surfaced pens of glass, concrete, or sheet steel. Automatic watering valves and other projections should be avoided. The dogs' nails should be kept trimmed. Some dogs might have to wear Elizabethan collars for extended periods to prevent self-inflicted wounds. Dogs should be given opportunities for exercise, either singly or in small groups, by being released under supervision into an exercise pen or room that is free of sharp projections. Leash-walking should be avoided.

Wound management is relatively simple. Wounds tend to heal well, possibly because hyperextensible skin places little tension on wounds. Cutting suture needles and single-stranded nylon suture material can tear through the skin, but tapered needles and braided sutures, such as those of polygalactin 910 (Vicryl), are well tolerated. It is important to avoid placing too much tension along a single suture line. Hygromas and hematomas, which can become large under loose skin, can be encountered, either as sequelae to lacerations or as primary events. Adhesive tape should never be applied directly to the skin or fur during bandaging because it can tear the skin when the bandage is removed. Entropion and ectropion can be corrected surgically; however, repeated correction might be necessary.

Reproduction

All affected dogs appear to be heterozygotes; affected homozygotes probably die in utero. To increase fertility, to avoid injury of affected animals, and to prevent conception of homozygotes, it is preferable to select normal bitches and affected males for breeding and to use artificial insemination. Heterozygous affected pups can be identified at birth by the fragility and hyperextensibility of their skin, as can heterozygous fetuses in late gestation.

Endocrinologic Diseases

Clinical Features

Endocrinopathies in the dog pose diagnostic and therapeutic challenges because they are complicated physiologic derangements that often involve multiple organ systems. An endocrinopathy might be a desired element of an experimental design or simply a spontaneous random occurrence that would be expected in any canine population. Table 6.4 lists the major endocrinopathies that have been documented in dogs. Discussions in this section are limited to endocrinopathies that either are induced in experimental animals or are undesired results of management procedures or investigational protocols. Hypothyroidism and hyperadrenocorticism (Cushing's disease), two major endocrinopathies often seen in clinical veterinary practice, are not discussed here but are well described in the veterinary medical literature (e.g., Capen and Martin, 1989; Chester, 1987, Drazner, 1987a; Feldman, 1989; Hsu and Crump, 1989; Peterson and Ferguson, 1989). A brief review of disorders of calcium metabolism is included because hypocalcemia caused by iatrogenic hypoparathyroidism occasionally occurs in a research setting, and hypercalcemia is often mistakenly attributed to parathyroid dysfunction.

TABLE 6.4 Selected Endocrine Disorders in Dogs

Affected Organ	Diseases
Adrenal cortex	Hyperadrenocorticism
	Hypoadrenocorticism
Adrenal medulla	Pheochromocytoma
Pancreas	Diabetes mellitus
	Gastrinoma
Parathyroid	Hyperparathyroidism
	Hypoparathyroidism
Pituitary	Acromegaly
	Diabetes insipidus
	Hypopituitarism
Thyroid	Hyperthyroidism
	Hypothyroidism
Multiple glands	Hyperlipidemia
	Hypoglycemia

TABLE 6.5 Common Clinical Signs of Selected Canine Endocrinopathies

Endocrinopathy	Common Clinical Signs
Diabetes mellitus	Hyperglycemia, polydipsia, polyuria, glycosuria, increased food consumption but loss of weight, bilateral cataract development, weakness
Hypoadrenocorticism	Weakness, vomiting, diarrhea, bradycardia, acute collapse
Acromegaly	Respiratory stridor, increased interdental spaces, prominent skin folds, abdominal enlargement, fatigue
Hypercalcemia	Mental dullness; muscular weakness; tachycardia; upper gastrointestinal signs, including anorexia, nausea, and vomiting; signs of renal disease, including nephrocalcinosis, renal calculi, and secondary renal failure
Hypocalcemia	Muscle tremors, tetany, seizures

Common clinical signs of the endocrinopathies to be discussed are listed in Table 6.5. They range from very subtle changes to acute crises. Most are nonspecific and can also be seen in various nonendocrine disorders. Detailed discussions of endocrinopathies can be found in the veterinary medical literature (e.g., Drazner, 1987b; Ettinger, 1989; Feldman and Nelson, 1987; McDonald and Pineda, 1989; Morgan, 1992).

Husbandry and Veterinary Care

Procedures for managing dogs with endocrinopathies are dictated by both the experimental design and the animals' welfare.

Diabetes mellitus. Diabetes mellitus in the dog is a recognized spontaneously occurring model (Kramer, 1981), and the disease is readily induced either by chemical ablation of the pancreatic β-cells or by total pancreatectomy (Mordes and Rossini, 1985). Frequent monitoring is mandatory for the successful management of dogs with diabetes mellitus. Daily measurements, before the first meal of the day and 6-12 hours later, are required to stabilize and control blood glucose in diabetic dogs. The second glucose measurement can be eliminated only when the afternoon blood glucose of an individual dog is consistent from day to day and the insulin requirement for that dog is well established. Blood glucose monitoring should begin after initial administration of diabetogenic chemicals or during the first 24 hours after pancreatectomy. Fasting blood glucose, as measured by the plasma or serum glucose oxidase method, ranges from 65-118 mg/dL (3.6-6.5 mmol/L) in normal adult dogs (Kaneko, 1989).

A number of insulin preparations can be used either singly or in combination in dogs: regular, NPH, lente, and ultralente. Unit doses and preparation types must be determined for and adjusted to the response of each dog. Insulin should be started at a dose of 1 U/kg per day injected subcutaneously at the time of feeding the first meal of the day. Daily proportions of each preparation included in a therapeutic regimen are determined by trial and error as guided by the results of serial blood glucose measurements. Detailed information on dosage and characteristics of various insulin preparations is available (Nelson, 1989; Schaer, 1992).

In addition to insulin administration, stresses from environmental and experimental manipulation, exercise, concurrent disease, estrus, and changes in food and water intake can cause profound fluctuations in blood glucose concentrations. Blood glucose can be manipulated by adjusting insulin types and dosages. As a general rule, it is preferable to have a slightly hyperglycemic dog rather than a hypoglycemic one because of the potentially disastrous results of a hypoglycemic crisis. If such a crisis occurs, it should be treated with intravenous dextrose and supportive care (Kirk and Bistner, 1985). Supplemental glucose can be given orally if the dog is able to swallow. Obviously, a necessary follow-up includes reviewing and adjusting the insulin dosage and the ratio of short- to long-acting insulin preparations given.

The amount of food fed to each diabetic dog should be standardized at what is necessary to maintain its optimal body weight. The same amount should be fed each day. Once an eating pattern (amount of food eaten and time required for meal consumption) is established for a given dog, its appetite can be used as an indicator of general well-being.

In pancreatectomized dogs, it is necessary to compensate for lost pancreatic exocrine function. That can be accomplished by adding a commercially available digestive enzyme to the food. Some dogs find the product unpalatable, but it is generally accepted if it is mixed with canned food.

Diabetic dogs can be maintained for long periods, but sequelae of diabetes mellitus—including neuropathy, immune system compromise, and delayed healing—do occur, and a shorter than normal life span should be expected.

Hypoadrenocorticism. The canine model of hypoadrenocorticism (Addison's disease) is a classic model in biomedical research (Brown-Sequard, 1856). Hypoadrenocorticism can be induced in dogs by administering the drug mitotane,[1] which chemically ablates the adrenal cortex (Nelson and Woodard, 1949). During induction, a presumptive diagnosis can be made by monitoring changes in serum electrolytes, specifically sodium and potassium. The normal ranges of sodium and potassium concentrations in dog serum are

---

[1] Chemical name, 1,1-dichloro-2-(*o*-chlorophenyl)-2-(*p*-chlorophenyl) ethane; trivial name, *o,p'*-DDD; brand name, Lysodren.

140-155 mEq/L and 3.7-5.8 mEq/L, respectively (Carlson, 1989). In dogs with hypoadrenocorticism, the sodium-to-potassium ratio is decreased to less than 27:1 (Schrader, 1988), although this hyperkalemia is not pathognomonic. The adrenal corticotropic hormone stimulation test is required for definitive diagnosis (Nichols and Peterson, 1992). In a crisis, resuscitation requires recognizing the problem, intravenously administering 0.9 percent saline solution, replacing glucocorticoids and mineralocorticoids, and possibly providing therapy for hyperkalemia. Long-term maintenance entails glucocorticoid (cortisone) administration, mineralocorticoid supplementation with 9-fluorohydrocortisone acetate,[2] and the addition of sodium chloride to the diet. Electrolytes should be monitored at least weekly once stabilization is achieved. Environmental and experimental stresses and alterations in water and food availability can have substantial effects on electrolyte balance and homeostasis. Additional glucocorticoid (increased by a factor of 2-10) should be administered during periods of stress.

Acromegaly. Acromegaly can be iatrogenically induced in bitches when progesterone is given to prevent estrous cycling (Eigenmann, 1985, 1989). It can also be secondary to increased production of progesterone during diestrus. Progesterone induces acromegaly by increasing the production of growth hormone in the anterior pituitary gland. The excessive release of growth hormone can also induce a "pituitary diabetes" that can be difficult to control with insulin. Cessation of progesterone administration or spaying will reverse acromegalic changes.

Calcium derangements. Although disorders of the parathyroid glands are usually suspected when hypercalcemia or hypocalcemia is present, the calcium abnormality is more often associated with other conditions, including pseudohyperparathyroidism, the most common cause of hypercalcemia (Feldman and Nelson, 1987); hypoadrenocorticism; renal failure; bone lesions; and hypervitaminosis D. Primary hyperparathyroidism in the dog is rare. Pseudohyperparathyroidism (hypercalcemia of malignancy) is a paraneoplastic syndrome that has been recognized in dogs with lymphosarcoma, adenocarcinoma of the anal apocrine glands, multiple myeloma, osteosarcoma, and other neoplasms (Meuten et al., 1982, 1986). Signs of hypercalcemia are not always overt, and treatment should be directed toward the underlying cause.

Causes of hypocalcemia include calcium imbalance during lactation, renal disease, acute pancreatitis, intestinal malabsorption, hypoalbuminemia, and primary hypoparathyroidism (idiopathic or iatrogenic). Iatrogenic

---

[2] Brand name, Florinef.

hypoparathyroidism is associated with inadvertent damage or removal of the parathyroid glands and is an important consideration in research settings. Surgery involving the ventral neck area or the laryngeal-tracheal area or removal of the thyroid glands carries an increased risk of complications related to parathyroid function. Treatment includes calcium replacement and appropriate management of the precipitating disorder.

Hematologic Disorders

Clinical Features

Canine models of human hematologic disorders have been reviewed (Dodds, 1988, 1989, 1992; Hall and Giger, 1992; Harvey, 1989; Kaneko, 1987; Knoll, 1992). Clinical signs of some of these disorders are listed in Table 6.6.

TABLE 6.6   Inheritance and Signs of Selected Hematologic Disorders in Dogs

Disorder	Inheritance	Clinical signs
Hemophilia A	X-linked	Low factor VIII coagulant activity but normal or increased von Willebrand factor antigen concentrations; spontaneous bleeding diathesis of varied severity, depending on factor VIII activity; severely affected dogs often exhibit spontaneous hemarthroses and large joints. The most common severe inherited bleeding disease. Recognized in most purebreds and in mongrels.
Hemophilia B (Christmas disease)	X-linked	Deficiency of factor IX activity; signs similar to those of hemophilia A. Recognized in 17 breeds.
von Willebrand's disease type I	Autosomal incompletely dominant	Variable deficiency of von Willebrand factor; factor VIII activity might be reduced; and prolonged bleeding time; moderately severe bleeding diathesis of mucosal surfaces. Signs are exacerbated by stress, hypothyroidism, intercurrent disease, trauma, and surgery. Recognized in more than 50 breeds.

continued on next page

TABLE 6.6   Continued

Disorder	Inheritance	Clinical signs
von Willebrand's disease type III	Autosomal recessive	Severe deficiency of von Willebrand factor; factor VIII activity is usually low; indefinitely prolonged bleeding time; mucosal surface bleeding diathesis, which can be severe and is exacerbated by stress, hypothyroidism, trauma, surgery, and intercurrent disease. Recognized in Chesapeake Bay retrievers, Scottish terriers, and Shetland sheepdogs.
Factor X deficiency	Autosomal incompletely dominant	Homozygotes are stillborn or die shortly after birth; affected pups might live for up to 2 weeks and then die of massive internal bleeding; young adults can also exhibit life-threatening hemorrhage, but signs in mature adults are usually mild and confined to mucosal surfaces. Found only in one large family of cocker spaniels.
Thrombopathia	Autosomal	Affected dogs can have no clinical signs or show increased bleeding tendency that can be exacerbated by trauma or surgery. Found in basset hounds and otterhounds.
Cyclic hematopoiesis	Autosomal recessive	Regularly occurring interruptions of bone marrow hematopoiesis with loss of neutrophils from peripheral blood; during these periods, dogs exhibit fever, enteritis, keratitis, pneumonia, and skin infections; infections can become life-threatening if not treated. Found in gray collies.
Pyruvate kinase deficiency	Autosomal recessive	Affected dogs exhibit severe anemia with reticulocytosis, macrocytosis, and polychromasia; hyperbilirubinemia; splenomegaly with extramedullary hematopoiesis; and decreased red cell survival. Found in basenjis, beagles, and cairn terriers.
Erythrocyte phosphofructokinase deficiency	Autosomal recessive	Persistent compensated hemolytic anemia with episodes of intravascular hemolysis, hemoglobinuria, and fever associated with stress or exercise; hemolytic crises follow hyperventilation-induced alkalemia; red cells of affected dogs are extremely alkaline and fragile in vitro. Found in English springer spaniels.

Husbandry and Veterinary Care

Bleeding disorders. Dogs with congenital and acquired bleeding disorders require special housing to minimize the risk of spontaneous or injury-induced bleeding. This is important not only for the animals' welfare, but also for experimental reasons. The basal state of animals that experience repetitive bleeding can be altered by the physiologic stress that such bleeding causes and, if bleeding is severe enough to require transfusions, by repeated exposure to homologous plasma proteins and blood cells. That is of particular concern for dogs with severe disorders, such as hemophilia.

Dogs with bleeding disorders should be housed in enclosures that have smooth sides and fronts with smooth vertical or cross-hatched bars. It is not advisable to use materials that can be climbed (e.g., chain-link fencing) because dogs with bleeding disorders can suffer foot injuries caused by weight-bearing pressure between the toes. Enclosure size is also important. To prevent injury, affected animals should have sufficient space to move about freely but not enough to permit vigorous exercise if they become excited. Enclosures should be square or oblong; injury is more likely to occur in a long, narrow run, especially in dogs with long tails, which during wagging can be traumatized by hitting against the sides. Experience has shown that for dogs weighing from 13.6-36.3 kg (30-80 lb), primary housing measuring about 4 × 6 ft (1.22 × 1.83 m) or 5 × 5 ft (1.52 × 1.52 m) minimizes the risk of injury.

Severely affected dogs should be housed individually because the risk of injury in playing with other dogs is substantial. To provide socialization, it is advisable to construct pens that allow visual contact between dogs; this can be achieved by building pens across an aisle from or perpendicular to each other. Partitions between the runs should be solid for the first 4 ft (1.22 m) in height to prevent injury caused by dogs in adjacent pens playing or fighting through the partition, and the seam with the floor should be smooth.

To avoid foot-pad abrasions, nonslip flooring should not be too rough. A poured rubberized flooring with a small amount of sand added to the last coat should create enough friction to prevent sliding. Nontoxic bedding (e.g., shredded newspaper or shavings) can be used to minimize injuries if sliding does occur. English rubber coits or tennis balls can be used to provide environmental enrichment.

Special arrangements are required for feeding and watering. Automatic watering devices are generally not recommended because the spigots can cause mouth injuries, and bleeding from such injuries is usually difficult to control. It is better to use large water buckets anchored to the sides or fronts of the pens. Dry food should be softened before feeding and supplemented with good-quality canned or cooked meat. A hematinic can be added to the food for conditioning. Hard biscuits should not be fed.

Bleeding from small surface injuries to the gums or nose or from toenails that are cut too short can be stopped by using sealant materials, such as Nexaband glue (Tri-Point Medical, LP, Raleigh, N.C.). Bleeding toenails can also be packed with styptic powder, and the soft rubber end cap from an intravenous set or catheter can be wedged tightly over the nail. If necessary, the foot can be bandaged; this supplies enough local pressure to control the bleeding. For animals that experience severe bleeding episodes, transfusion is the treatment of choice. Fresh-frozen plasma, plasma concentrates, platelet concentrates, or packed red cells should be given as required for the specific disorder. Details of management and treatment are summarized elsewhere (Dodds, 1989, 1992).

Another management procedure to keep animals healthy and reduce bleeding risk is prophylactic dentistry, which must be performed very carefully to avoid injury to the gums. Booster vaccinations should not be given during bleeding episodes because they create a transient platelet deficit (Dodds, 1992). In addition, dogs are at increased risk for bleeding episodes for 10-14 days after vaccinations. Affected females sometimes bleed excessively both during estrus and during the 30-40 days beforehand when estrogen concentrations are elevated.

Cyclic hematopoiesis. Colonies of grey collies with cyclic hematopoiesis (formerly called cyclic neutropenia) have special requirements because they are susceptible to recurring infections and anemia (Knoll, 1992). They have a cyclic, profound drop in all their blood-cell classes, although the numbers of each cell type rise and fall at different times. Affected animals rarely live beyond the age of 3 years and experience frequent bleeding episodes from cyclic thrombocytopenia. Respiratory tract and enteric infections are the most debilitating.

Affected animals can often be housed together, but they need scrupulously clean facilities to minimize infection, close clinical monitoring, and supportive therapy. They should be monitored for neutropenia, and prophylactic antibiotics should be administered as neutrophil counts begin to decline.

Other hematologic disorders. Dogs with various other inherited and acquired hematologic diseases also require special care. For example, basenjis with pyruvate kinase deficiency and recurring anemia must be closely monitored because of their increased susceptibility to infection or stress (Hall and Giger, 1992; Harvey, 1989); beagles with hereditary nonspherocytic hemolytic anemia must be closely monitored for episodes of hemolytic crisis (Maggio-Price et al., 1988); and English springer spaniels with erythrocyte phosphofructokinase deficiency require special care during episodes of hemoglobinuria or myoglobinuria (Hall and Giger, 1992; Harvey, 1989).

Reproduction

For dogs with severe inherited bleeding disorders—such as hemophilia, von Willebrand's disease, factor X deficiency, and platelet dysfunction (thrombopathia)—special care is needed for breeding, whelping, and rearing of the offspring. Immediately after birth, each pup should be carefully examined for signs of bleeding, its umbilical cord should be ligated, and the potential for trauma from the dam should be minimized. It might be necessary to tranquilize first-time dams slightly to protect the young. When the pups are weaned and start to become more active, blood samples should be taken to determine which pups are affected. In hemophilia, the affected pups from a carrier (heterozygous) dam will be males, unless the sire is a hemophiliac (hemizygote), in which case both affected hemizygote males and homozygote females can be produced. Generally, male pups should be watched more closely, and the affected ones should be removed and housed separately if the litter is too rambunctious. Cages should be relatively small; a floor area of about 30×36 in (76×91 cm) is recommended for the average hemophilic pup.

Affected pups should be watched carefully after vaccinations. Modified live-virus vaccines might induce a relative thrombocytopenia and platelet dysfunction during the period of viremia (i.e., 3-10 days after vaccination) (Dodds, 1992). The pups are at substantial risk for spontaneous or traumatic bleeding at this time because the vaccine effect on platelet function superimposes another hemostatic burden. All vaccinations should be given subcutaneously with a small-gauge needle, preferably 23 or 25 gauge, in the loose skin folds of the neck. Intramuscular injections in affected animals should be avoided.

Affected pups should be housed initially in cages and eventually in small pens. At teething, affected puppies often bleed excessively from the gums; this necessitates use of a topically applied sealant and, on occasion, transfusion therapy.

Immunologic Diseases

Primary Immunodeficiency and Autoimmune Diseases

Clinical Features

Immunodeficiency is characterized by failure to manifest a normal immune response when challenged by infectious agents or other substances that are foreign to the body. The subnormal response can result from a defect in the afferent, central, or efferent limb of the immune system (see review in NRC, 1989). Immunodeficiency disorders can be primary (i.e., inherited) or secondary (i.e., acquired). Primary immune deficiency can result from an inherited defect in immunocompetent cells or effector mechanisms (e.g., complement or phagocytes) or can be associated with autoimmune disease or a deficiency in growth factors necessary for the optimal function of immunocompetent cells (WHO Scientific Group, 1986). Secondary immune deficiency can be caused by various environmental factors, including x rays, viral agents, toxic chemicals, and dietary deficiencies.

Several primary immunodeficiency diseases have been described in dogs, including selective IgA deficiency (Campbell, 1991; Felsburg et al., 1985; Moroff et al., 1986), IgM deficiency (Mill and Campbell, 1992; Plechner, 1979), common variable immunodeficiency (A. Rivas, New York State College of Veterinary Medicine, Cornell University, Ithaca, N.Y., unpublished), and severe combined immunodeficiency disease (Jezyk et al., 1989; Patterson et al., 1982). Dogs with particular autoimmune diseases also suffer from immunodeficiency. A high incidence of septicemia has been observed in dogs that were bred to develop systemic lupus erythematosus (SLE) (Quimby et al., 1979). Autoimmune hemolytic anemia (AHA) (Bull et al., 1971; Dodds, 1983; Klag et al., 1993), immune thrombocytopenic purpura (ITP) (Dodds, 1983, 1992; Waye, 1960), SLE (Grindem and Johnson, 1983; Monier et al., 1988; Quimby, 1981), rheumatoid arthritis (RA) (Bell et al., 1991; Carter et al., 1989; Quimby et al., 1978), Sjögren's syndrome (Kaswan et al., 1985; Quimby et al., 1979), autoimmune thyroiditis (Gosselin et al., 1982; Quimby et al., 1979; Rajatanavin et al., 1989; Thacker et al., 1992), and thyrogastric disease (Quimby et al., 1978) have been found in research dogs. Primary immunodeficiencies in dogs have also been associated with the absence of the third component of complement (Winkelstein et al., 1981); deficits in neutrophil function, including cyclic hematopoiesis (see page 100) (Knoll, 1992; Lund et al., 1967) and granulocytopathy (Knoll, 1992; Renshaw and Davis, 1979); dysregulation of interleukin-6 (DiBartola et al., 1990; Rivas et al., 1992); and deficiency of growth hormone (Roth et al., 1980). Clinical signs of these diseases are presented in Table 6.7.

All dogs with primary immunodeficiencies are predisposed to infection. Dogs with disorders associated primarily with hypogammmaglobulinemia, complement, or phagocytic function are predisposed to bacterial infection (Blum et al., 1985; Lund et al., 1967; Moroff et al., 1986; Renshaw and Davis, 1979). Those with disorders of cell-mediated immunity have increased susceptibility to fungi and viruses (Jezyk et al., 1989).

TABLE 6.7 Clinical Signs of Selected Primary Immunodeficiency and Autoimmune Diseases in Dogs

Immunologic Disease	Clinical Signs
Common variable immunodeficiency	Increased susceptibility to infectious diseases; clinical presentation after the age of 6 months
IgM deficiency	Increased susceptibility to bacterial diseases
Selective IgA deficiency	Increased susceptibility of some dogs to infectious diseases of mucosal surfaces, such as those of gastrointestinal, respiratory, and urogenital tracts
Severe combined immunodeficiency	Extreme susceptibility to bacterial, viral, and fungal infections; clinical presentation in first few weeks of life; death before reaching maturity
Autoimmune hemolytic anemia	Pallor, slight jaundice, splenomegaly, lymphadenopathy, weakness, and shortness of breath; profound anemia and recurrent episodes of hemolytic disease in approximately 50% of affected dogs
Immune thrombocytopenic purpura	Bruise easily, prolonged bleeding after trauma
Systemic lupus erythematosus	Rash, hemolytic anemia, immunothrombocytopenic purpura, polyarthritis, and proteinuria; females affected more frequently than males
Rheumatoid arthritis	Swollen painful joints—generally multiple small articular joints
Sjögren's syndrome	Keratoconjunctivitis sicca (dry eyes); corneal ulcers associated with dry eyes; excessive dental caries; inflamed gums; signs associated with hypothyroidism, including tendency to obesity, tendency to seek warm places, bilaterally symmetrical hair loss, and changes in skin thickness
Autoimmune (lymphocytic) thyroiditis	Signs associated with hypothyroidism, including tendency to obesity, tendency to seek warm places, bilaterally symmetrical hair loss, and changes in skin thickness
Thyrogastric disease	Signs associated with hypothyroidism, inappetence, megaloblastic anemia, and atrophic gastritis
Granulocytopathy	Increased susceptibility to bacterial infections
Dysregulation of interleukin-6	Familial Mediterranean fever, characterized by fever, synovitis, and renal failure
Deficiency of growth hormone	Small body stature; generalized increase in susceptibility to infectious diseases

Husbandry and Veterinary Care

Immunodeficient dogs pose special management problems. Immune diseases must be diagnosed, their prognosis determined, and their therapy monitored. A number of tests have been developed for those purposes, including tests that assay T- and B-cell function (Ladiges et al., 1988, 1989), identify serologic markers of autoimmune diseases (Kaplan and Quimby, 1983; Quimby et al., 1980), identify circulating immune complexes in rheumatic and neoplastic diseases (Carter et al., 1989; Terman et al., 1979), and assay phagocyte function (Smith and Lumsden, 1983).

The susceptibility of immunodeficient dogs to infectious diseases is handled in various ways. All immunodeficient dogs can benefit from an environment that minimizes contact with canine pathogens; however, for some of these conditions (e.g., severe combined immunodeficiency), cesarean derivation and maintenance in a gnotobiotic chamber are required to ensure survival. Pups with humoral deficiencies born to normal dams profit from receiving maternal antibodies in colostrum, and their dams should be immunized before being bred to ensure that high concentrations of antibodies will be present. Adult dogs with humoral deficiencies can be helped by transfusions of normal or hyperimmune serum or plasma or by administration of purified gamma globulin. Some dogs that are genetically predisposed to autoimmune diseases can be spared clinical illness for years by housing them in gnotobiotic chambers; however, if they are moved to a conventional environment, they quickly develop autoimmune disease (Schwartz et al., 1978).

Dogs with autoimmune diseases should be carefully monitored and appropriately treated. Treatment might involve immunosuppressive therapy (e.g., for SLE, AHA, ITP, and RA), transfusions of red cells and platelets (for AHA and ITP), splenectomy (for AHA and ITP), renal dialysis (for SLE), administration of thyroxine (for autoimmune thyroiditis), administration of thyroxine and vitamin B_{12} (for thyrogastric disease), and administration of artificial tears (see the section on ophthalmologic disorders) and special dental care (for Sjögren's syndrome). Some dogs with growth hormone deficiency benefit from injections of thymosin (Roth et al., 1980). Bone marrow transplantation and systemic antibiotics are effective in treating dogs with neutrophil defects. Dogs with thrombocytopenia (as in SLE, ITP, or Evan's syndrome) are predisposed to bleeding and bruising and should be housed and maintained as described in the section on hematologic disorders. Preliminary studies suggest that oral levamisole therapy is efficacious in treating one type of canine common variable immunodeficiency that is associated with ulcerative colitis and a predisposition to adenocarcinoma of the intestine (A. Rivas, New York State College of Veterinary Medicine, Cornell University, Ithaca, N.Y., unpublished). Trials involving the use of colchicine to delay the onset of amyloidosis in dogs with interleukin-6 dysregulation are in progress (L. Tintle, Wurtsboro Veterinary Hospital, Wurtsboro, N.Y., unpublished). The care of dogs with C3 deficiency and dogs that have been exposed to total body irradiation and immunosuppressive drugs associated with organ transplantation is described below. Dogs should be immunized against known canine pathogens before being exposed to agents that will induce immunodeficiency.

Reproduction

In colonies where the objective is to reproduce dogs with SLE by selecting breeders with serologic evidence of the disorder (i.e., by using antinuclear antibody and LE-cell tests), many progeny develop autoimmune diseases not apparent in the parents (Monier et al., 1988; Quimby et al., 1979). That observation has led to the hypothesis that multiple genes control the susceptibility and specificity of autoimmune diseases (Monier et al., 1988; Quimby and Schwartz, 1978). In some cases, an unanticipated result is compromised fertility (e.g., immune-mediated aspermatogenesis), which necessitates the use of littermates or repeat breeding of the parents to continue the lineage (Quimby et al., 1978). Hypothyroidism caused by lymphocytic thyroiditis (Beierwaltes and Nishiyama, 1968; Gosselin et al., 1982; Mizejewski et al., 1971; Rajatanavin et al., 1989; Thacker et al., 1992) can lead to poor reproductive performance that can be corrected with thyroxine-replacement therapy. Details on monitoring blood thyroxine and oral supplementation have been published (DePaolo and Masoro, 1989; Ferguson, 1986). For some autoimmune diseases, such as immune-mediated aspermatogenesis, no therapy has been found.

Complement Deficiency

Clinical Features

Dogs deficient in the third component of complement (C3) are particularly susceptible to bacterial infections (Blum et al., 1985). They also develop a membranoproliferative glomerulonephritis, which can be detected histologically by the age of 1 year (Cork et al., 1991). Affected dogs are normally active and appear well; the only clinical sign of this renal disease is proteinuria. Renal disease progresses inexorably and culminates in a nephrotic syndrome with azotemia when the dogs are 6-8 years old.

Husbandry and Veterinary Care

Dogs deficient in C3 can be reared and housed in standard laboratory dog facilities. Because the dogs are susceptible to bacterial infections (Chick et al., 1984), animal technicians should be alert to any deviations from normal behavior that might indicate illness (e.g., inappetence and lethargy). C3-deficient dogs that show these clinical signs must immediately be evaluated for increased body temperature and leukocytosis. Blood samples should be taken and submitted for culturing to identify and determine the antibiotic sensitivity of the microorganisms; however, treatment with intravenous bactericidal antibiotics should not await diagnosis but should begin as soon as clinical signs are detected and a blood sample has been drawn. Although that protocol undoubtedly results in overtreating and might preclude a definitive diagnosis, it will in most cases ensure the recovery and survival of the affected dog. If an invasive procedure (e.g., renal biopsy or placement of an indwelling catheter) is required, antibiotic prophylaxis should begin 24 hours beforehand, and it is essential to follow strict aseptic technique while performing the procedure.

The presence of proteinuria can be detected by testing for total-protein excretion in the urine over a 24-hour period, and renal biopsies can be used to evaluate the progression of renal disease. As dogs age, periodic measurements of total serum protein, albumin, and serum urea nitrogen can be used to identify dogs whose renal disease is becoming severe or those in which a nephrotic syndrome might lead to fluid accumulation in body cavities. Repeated blood transfusions or infusions of canine plasma are contraindicated because they exacerbate renal disease.

Reproduction

C3 deficiency is inherited as an autosomal recessive trait (Johnson et al., 1986; Winkelstein et al., 1982). Affected pups are produced by breeding heterozygous females with homozygous males. Homozygous females are fertile but have rarely produced viable young. Pups should be tested at birth, and the ones that are C3-deficient should be placed on antibiotic therapy for the first 4 days after birth. C3-deficient dogs do not respond normally to immunization; therefore, it is recommended that immunizations against the common canine pathogens be given at 2-week intervals until the pups are 18 weeks old (Krakowka et al., 1987; O'Neil et al., 1988; Winkelstein et al., 1986).

Organ Transplantation

Clinical Features

Dogs that are used in organ-transplantation studies must first be made immunodeficient. Immunosuppressive methods include total-body irradiation and administration of cytotoxic chemicals (Ladiges et al., 1989). Immunosuppressed dogs are very susceptible to infectious diseases and might have gastrointestinal tract problems.

Husbandry and Veterinary Care

Dogs that undergo experimental organ transplantation generally require intensive postoperative supportive care, the level of which depends on the transplantation procedure used and the degree of immunosuppression required to overcome graft rejection. Supportive care includes fluid therapy, blood and platelet transfusions, preoperative and postoperative administration of appropriate antibiotics, and intensive husbandry practices. Regular monitoring of white cells is critical for ascertaining health status and determining the necessity for treatment. Blood should be cultured if clinical signs suggest septicemia. Nutritional needs are critical for dogs undergoing bowel transplantation or for those suffering from gastrointestinal tract problems caused by the immunosuppressive procedures. Dogs might need to be housed individually in intensive-care facilities during early convalescence.

Dogs undergoing bone marrow transplantation are profoundly immunodeficient for 200-300 days after lethal total-body irradiation and successful marrow engraftment, and they require intensive supportive care (Ladiges et al., 1990). Recovery of granulocyte count and function is complete by the twenty-fifth day after engraftment; blood lymphocyte count does not return to normal until day 200. Antibody response to bacteriophage and sheep and chicken red cells is lower than normal during the first 200 days, with IgM being the primary isotype. Lymphocyte stimulation by phytohemagglutinin, the mixed-leukocyte reaction, and the response to first- and second-set skin grafts are impaired. Long-term survivors (dogs that survive more than 200 days) generally regain their health and are no longer more susceptible than normal to infectious diseases. The development of graft-versus-host disease and its treatment drastically affect recovery of the immune system and place the dogs at increased risk for contracting infections.

Lysosomal Storage Diseases

Clinical Features

Clinical manifestations of canine lysosomal storage diseases (LSDs) generally fall into three categories: severe neurologic signs, mainly skeletal signs, and a mixture of visceral, skeletal, and neurologic signs. The following discussion addresses techniques for managing dogs in each category, using a single LSD as an example. The techniques can be extended to manage dogs with other LSDs.

Fucosidosis. Fucosidosis is caused by a deficiency of α-L-fucosidase (Healy et al., 1984). Affected dogs exhibit mainly neurologic signs. By the age of 12 months, affected dogs show subtle behavioral changes and might have an overextended posture. From 12 to 18 months, they develop mild ataxia and hypermetria. Signs progress rapidly between the ages of 18 and 24 months to more severe deficits in gait, proprioceptive defects, hyperclonus, nystagmus, kyphosis, and a loss of learned behavior. The dogs become dull and unresponsive. Hearing and vision might be impaired. Signs in severely affected, 24- to 36-month-old dogs include severe incoordination, opisthotonos, muscle spasms, muscle wasting, circling, head tilt, abnormal pupillary light reflexes, dysphagia, and cranial nerve deficits. The dogs become severely obtunded and suffer from self-inflicted injury. If not euthanatized, they usually die by the age of 3 years.

Mucopolysaccharidosis VII. The majority of clinical signs in canine mucopolysaccharidosis VII (MPS VII), a condition caused by a deficiency of β-glucuronidase, are related to skeletal and joint abnormalities (Haskins et al., 1984). Progressive noninflammatory arthrosis develops, and joints become lax and deformed. By the age of 3-6 months, affected dogs are unable to stand, and the muscles of locomotion atrophy. Corneal clouding can lead to decreased vision in dogs with MPS VII, but the impairment is generally less severe than in dogs with MPS I. At the age of 15-22 months, MPS VII-affected dogs often become dull and lethargic and lose interest in their environment and in animal-care personnel. Those signs might be associated with progressive hydrocephalus.

Mucopolysaccharidosis I. Canine mucopolysaccharidosis I (MPS I), a condition caused by a deficiency of α-L-iduronidase, is most similar to the human MPS I phenotype of intermediate severity (Hurler's syndrome and Scheie's syndrome) (Shull et al., 1982). Clinical signs refer to visceral, skeletal, and mild neurologic injury. Dogs with MPS I appear normal at birth, although there is a higher than normal incidence of umbilical hernias. Affected pups remain generally healthy for 4-6 months and then show stunted growth, corneal clouding, and progressive, degenerative, noninflammatory joint disease caused by mucopolysaccharide deposition in synovial and periarticular tissues. Joint laxity caused by abnormalities in ligaments and tendons is also common and, in combination with the arthroses, causes decreased ambulation. Degeneration of intervertebral disks, collapse of disk spaces, vertebral and long-bone osteopenia, and spondylosis also develop. Mucopolysaccharide accumulation in heart valves and coronary arteries can cause rapidly progressing heart failure. Affected dogs remain alert and responsive until their death by natural causes or euthanasia, often between the ages of 2 and 3 years.

Husbandry and Veterinary Care

Dogs with LSD present unique and serious medical and husbandry problems. Proper care of these valuable, critically ill animal models requires compassion, diligence, hard work, and specialized knowledge of the diseases involved. Technicians must be well trained and observant.

Fucosidosis. As the clinical signs progress, affected dogs should be handled carefully to prevent injury. They should be fed, exercised, and housed separately from normal dogs. Severely ill dogs should be moved by carrying. Affected dogs should always be housed on a raised trampoline bed and kept dry during cage cleaning to prevent self-soiling and pressure sores. Particular attention should be given to dogs with long hair; they should be bathed weekly and clipped several times a year. Ears should be checked daily for signs of infection. Dogs with moderate to advanced disease should be fed more frequently, and canned or moistened dry food should be used to aid prehension. Dogs with advanced disease often have a poor appetite, and the addition of highly palatable foods assists in maintaining body weight. Excess dental tartar must be removed regularly. At the age of 2-3 years, motor and mental impairment will have progressed to the point that euthanasia will be indicated.

Mucopolysaccharidosis VII. MPS VII-affected dogs should be housed in cages with floors of coated wire mesh; this aids sanitation and helps to prevent decubital sores. Once the dogs are unable to walk, food and water intake must be carefully managed. Recumbent animals will usually eat and drink if pans of food and water are placed on the cage floor; however, hand feeding might become necessary. Euthanasia should be considered when a dog's response to human attention begins to diminish.

Mucopolysaccharidosis I. Except for surgical correction of umbilical hernias, special care is not usually required for dogs with MPS I that are less than 1 year old. However, as the disease progresses and the vertebral column deteriorates, the dogs become extremely fragile, and especially gentle handling is necessary when working with them or moving them between cages. Acute disk herniation can occur with even very minor trauma or inappropriate handling. Once skeletal disease has developed, exercise must be limited, and affected dogs must be protected from more rambunctious colony members. Decubital sores are a frequent consequence of the increase in time spent lying down. Housing affected dogs on shredded newspaper or elevated wire mesh provides both comfort and better sanitation.

Appetite generally remains normal, although hand feeding or varying the diet might become necessary, especially in dogs with pronounced corneal clouding, impaired hearing, or the rare decrease in cerebral sensorium. Some dogs have enlarged tongues; however, prehension of food is generally not a problem. The teeth of dogs that are fed a diet composed mainly of canned food require periodic scaling of tartar.

MPS I-affected dogs are rarely maintained until they die naturally. By the age of 24-36 months, the symptoms of skeletal disease are generally so marked that euthanasia is indicated before debilitation becomes unacceptable.

Reproduction

Most LSDs can impair fertility in dogs. MPS I- and VII-affected males have sired litters by artificial insemination. Males with fucosidosis show copulatory behavior before they become severely uncoordinated, but they are infertile because of epididymal lesions, which probably impair spermatozoan capacitation. Females with fucosidosis are fertile but are very poor mothers; their pups usually must be fostered or hand-reared. Pups with LSDs are generally produced by breeding heterozygous carriers that are clinically normal.

Muscular Dystrophy

Clinical Features

A genetic disorder homologous to Duchenne's muscular dystrophy of humans—a devastating, fatal disorder predominantly of boys—occurs in various breeds of dogs. The disorder in dogs, which is inherited as a simple sex-linked recessive gene with full penetrance, is known as canine X-linked muscular dystrophy, and dogs with the condition are called *xmd* dogs. The mutation has been found in golden retrievers and rottweilers, and a similar mutation is suspected to have occurred in samoyeds, malamutes, and Irish terriers. The golden retriever is the best studied of the affected breeds, and the following discussion is based on data on this breed.

Both Duchenne's muscular dystrophy and canine X-linked muscular dystrophy are caused by a defect in the production of dystrophin, a skeletal muscle cytoskeletal protein. The mutation in the dystrophin gene results in massive continuing skeletal muscle degeneration that occurs from birth onward. In dogs, progressive cardiac muscle degeneration begins in hemizygous males at the age of about 6 months. Carrier bitches appear clinically normal but have subtle lesions in their cardiac muscles. Because of the homology to Duchenne's muscular dystrophy, the *xmd* dog can serve as an animal model for studies leading to better understanding of the pathogenesis of Duchenne's muscular dystrophy, as well as for studies designed to assess therapeutic approaches (Valentine et al., 1992).

Clinical signs of obvious weakness, muscle wasting, and abnormal gait appear in *xmd* dogs at the age of about 8 weeks. After that time, clinical signs progress, and they are most severe at the age of about 6 months, at which time the dogs have a markedly stiff, shuffling gait. There is frequently a severely abnormal posture, with carpal overextension, tarsal overflexion, and splaying of the limbs. The dogs are unable to open their jaws fully, their tongues are thickened and cannot be fully extended, and they frequently drool excessively. After the age of 6 months, the clinical disease appears to stabilize, and many dogs seem to gain strength as they age. However, there is still a progressive degeneration and fibrosis of cardiac muscle that results in the characteristic Duchenne-type cardiomyopathy.

Husbandry and Veterinary Care

Dystrophic dogs do not require special caging. Shavings provide a soft, warm surface, but the shavings must be free of dust so that the dogs do not inhale particles and develop granulomatous pneumonia. Temperature and humidity must be carefully controlled. Older dystrophic dogs should be monitored carefully for signs of cardiac failure. Treatment for heart failure has been described (Fraser et al., 1991). Euthanasia should be considered when treatment fails to alleviate clinical signs (e.g., when the dog has difficulty breathing and when fluid accumulates in the abdomen).

Dystrophic dogs require high-calorie food that is easy to prehend and swallow because of the weakness of their tongue, jaw, and esophageal muscles. Canned food mixed with moistened dry food seems to constitute an adequate diet, but careful monitoring of food intake and weight is necessary. Regurgitation of food is common because of the esophageal skeletal muscle dysfunction. Severely disabled dogs might not be able to use automatic watering devices and might have to be given water in bowls or buckets. Their water might need frequent changing because of a buildup of saliva.

Adequate exercise is crucial during the period of rapid growth. Although dystrophic dogs might prefer to lie down, restricted exercise will result in more severe joint contractures. The presence of a slightly more active dystrophic cagemate is ideal, provided that competition for food does not impair food intake. The kennel must have a nonslippery surface to provide traction, and daily release for exercise is advised. These dogs should not be forced to exercise, however, because it might lead to increased muscle damage.

Dystrophic dogs cannot groom themselves adequately. Regular brushing of their haircoat and clipping of overgrown toenails is required. To prevent skin irritation, the mouth and jaw should be kept free of the saliva and food that accumulate.

Reproduction

Many dystrophic dogs survive to breeding age, and breeding colonies can be established. Some affected males are able to breed naturally; others are hampered by their physical disability and require artificial insemination techniques. An *xmd* male that breeds naturally might need assistance to remain upright once he has "tied" with the female. Breeding dystrophic bitches, which are produced by mating dystrophic males to carrier bitches, is possible but not advised. Pregnant dystrophic bitches require constant monitoring, are likely to have respiratory and cardiac complications, will require cesarean section, and might not be able to care for their pups adequately.

At whelping, a safe, warm environment and proper maternal care are essential for the survival of dystrophic pups. If dystrophic pups are stressed by cold, separation from the litter, or inability to compete with normal pups in a large litter, some of them will develop massive skeletal necrosis within the first few days of life. Once signs of severe weakness have developed in a pup, it is virtually impossible to save it. Severe diaphragmatic necrosis resulting in respiratory failure appears to be the cause of death. Dystrophic pups can be identified in the first week of life by their markedly increased serum concentrations of creatine kinase released from degenerating muscles. Dystrophic pups that survive the first week grow more slowly than their littermates. Euthanasia should be considered for pups that are too weak to nurse during the first week of life; tube feeding has not been successful in keeping such pups alive (B. A. Valentine, Department of Pathology, New York State College of Veterinary Medicine, Cornell University, Ithaca, N.Y., unpublished).

Neurologic Disorders

Clinical Features

Dogs with hereditary or induced neurologic disorders are often used to study equivalent human disorders. Clinical signs in these dogs include abnormal gait, hyperactivity, nervousness, tremors, convulsions, visual impairment, blindness, deafness, quadriplegia, and tetraplegia. Obviously, these dogs commonly require extra care to ensure that they are as comfortable as possible. Inherited canine neurologic diseases and their clinical signs have been reviewed by Cummings (1979) and Oliver and Lorenz

(1993); the pattern of inheritance of specific diseases has been discussed by Willis (1989).

Husbandry and Veterinary Care

Food and water must be placed where a neurologically impaired dog can find and reach them easily, and, if the dog is blind, placement should be consistent. That might require using water bowls instead of automatic waterers or, for dogs with severe impairment, intravenous or subcutaneous administration of fluids. Food might have to be placed in flat dishes, softened, or made into a gruel so that it can more easily be reached, masticated, and swallowed. Food and water intake should be monitored. Dogs should be weighed regularly to ensure that body weight is maintained. Nasogastric, lavage, pharyngotomy, or intragastric feeding might be required in some circumstances to provide adequate nutrition.

Dogs with sensory deficits can experience dysesthesias and might respond by chewing the affected limb or body part or another, more accessible body part. Several strategies can be used to deal with such behavior. Dogs should be closely monitored to detect the beginning of self-directed behaviors. A dog can sometimes be distracted by housing it where it has more external visual and social stimulation. If a nonaggressive cagemate can be identified, social housing might be sufficiently distracting—provided that the cagemate does not harass the affected dog or prevent it from eating and drinking. Toys, such as rawhide bones, might also be useful. If bandages must be used, they should not be too tight and should be checked regularly. Elizabethan collars or muzzles can be used to limit access to the body. Light tranquilization, if it does not interfere with the experimental protocol, might be helpful.

Dogs with sensory deficits might require extra or different bedding to prevent unintentional self-injury. The dogs' primary housing must be free of rough or sharp edges and projections. Dogs with motor deficits might have difficulty in positioning their bodies for urination and defecation. Sometimes all that is necessary is to provide flooring with better traction (e.g., plastic-coated grids or rubber mats). If necessary, research or animal-care personnel should assist the dog to position itself. Catheterization or manually expressing the bladder might be required to prevent urinary retention. Careful husbandry and nursing will avoid decubitus ulcers.

In dogs with respiratory deficits, the normal ability to thermoregulate by panting has been compromised. For these animals, exertion must be avoided and comfortable temperatures maintained.

Reproduction

Dogs with some neurologic disorders can reproduce, even though they are severely impaired. Such dogs usually need assistance for mating or require artificial insemination. Bitches with marked sensory or motor deficits or ataxia should be closely attended at parturition and while nursing to protect the pups from accidental injury. If the neurologic deficits of the dam interfere with her ability to care for her offspring, hand rearing or foster rearing will be required.

Ophthalmologic Disorders

Clinical Features

Dogs are affected by various ophthalmologic problems, either as inherent aspects of the research in which they are being used, as complications, or as acquired conditions unrelated to the research. Descriptions of canine eye diseases can be found in any standard text on veterinary ophthalmology (e.g., Gelatt, 1991; Helper, 1989). In the research setting, ocular problems that require special management techniques are visual impairment, painful ocular conditions, untoward sequelae of interfering with the eye's external protective mechanisms, and combinations of these conditions.

Blindness. Visual impairment in dogs usually cannot be measured precisely. For purposes of this discussion, *blindness* is used, in a loosely defined manner, to refer to any condition in which visual impairment is sufficient to interfere with a dog's ability to perform visually guided tasks or to exhibit normal visually guided behavior. In general, dogs maintained in a familiar environment adapt well to visual deficits that are congenital, are gradual in onset, or have been present for an extended time (weeks to months). A dog that has adapted to its blindness, that is maintained in a familiar environment, and that is not subjected to stressful experiences will move about actively and engage in all normal canine behavior. Its adaptation, or compensation, might be so successful that a naive observer will not recognize that it is blind.

Ocular pain. Painful ocular conditions fit broadly into three categories. External ocular pain is usually associated with corneal irritation and commonly causes obvious signs, such as blinking, excessive tearing, and redness. Uveal pain is caused by intraocular inflammation, which might not be evident without careful examination of the eye; uveal pain is usually more painful than corneal irritation. Glaucomatous pain is often the most insidious and most severe ocular pain. All these conditions are not only painful, but can threaten a dog's vision and the integrity of the affected eye.

Conditions associated with failure of the eye's external protective mechanisms. Untoward sequelae can arise from any condition that interferes with the eye's external defense mechanisms. These mechanisms depend on such funcitons as corneal sensitivity, lid movement, and tear production. Anything that reduces corneal sensation, interferes with lid movement, or lowers tear production can lead rapidly to painful ocular inflammation, impairment of vision, and loss of the affected eye. Common causes include anesthesia, radiation, surgical procedures, and drugs.

Husbandry and Veterinary Care

It is recommended that all experimental protocols involving dogs with ophthalmologic problems—whether the problems are "natural" (i.e., genetic), acquired, or induced—be reviewed by a veterinarian or a physician with training in ophthalmology (e.g., a veterinarian certified by the American College of Veterinary Ophthalmologists). Such protocols should include an adequate program for monitoring the dogs' ophthalmologic problems and written procedures for dealing with ocular emergencies.

Blindness. In spite of the ease with which dogs can adapt to blindness, they require special protection from a variety of environmental dangers, the more obvious of which are protruding objects, sharp edges, openings through which a dog might fall, and sources of electric or thermal injury. More insidious risks can arise because blind dogs lack the menace reflex, which normally protects the cornea from damage by causing the eyelids to blink in response to seen objects approaching the eye. Personnel responsible for the care and handling of blind dogs must be aware of these risks and keep them to a minimum and must watch for signs of acute or chronic corneal injury.

Dogs that have adapted to their blindness can become decompensated in response to rapid changes in their environment or other stressful experiences, such as anesthesia (e.g., for diagnostic, surgical, or experimental procedures), illness, and alterations in their daily routine. A decompensated chronically blind dog might look as though it has suddenly become blind and might exhibit behaviors compatible with a general stress reaction— from stiff-limbed hesitancy in walking and an apparent fear of its surroundings to anorexia or polydipsia, polyphagia, and polyuria. Similar signs can be observed in some dogs that have recently and rapidly lost their sight. Given time and a restricted, safe, and consistent environment, the blind dog will readapt and once again exhibit compensated normal behavior. Personnel responsible for the care and handling of blind dogs must be aware that these dogs need consistent familiar surroundings and that they might react adversely to stressful experiences. When approaching a blind dog, animal technicians should talk to it so that the dog will be more likely to perceive the approach as friendly.

Ocular pain. Ocular pain can vary from moderate to excruciating. Dogs in ophthalmologic research colonies are often at risk of developing ocular pain, either as a direct result of a study or as an unpredictable occasional side effect. In some cases, particularly if the pain is chronic or develops gradually, it will not be readily apparent without special examination procedures, especially if the observer is inexperienced. Personnel responsible for the care and handling of dogs used in ophthalmologic research should suspect that ocular pain is present when there is periocular soiling or when there are behavioral changes, such as decreased activity, decreased appetite, increased yawning, and changes in vocalization patterns.

Conditions associated with failure of the eye's external protective mechanisms. All protocols should be reviewed for potential adverse effects on external ocular defense mechanisms, and dogs subject to such risks should be monitored carefully for evidence of adverse effects.

Reproduction

Most dogs with ophthalmologic disorders can breed normally.

Orthopedic Disorders

Clinical Features

Dogs serve as models for both canine and human orthopedic diseases. Spontaneous bone and joint diseases of dogs have been reviewed (Lipowitz et al., 1993; Newton and Nunamaker, 1985; Whittick, 1990; Young, 1979). Orthopedic diseases can also be induced in dogs.

Husbandry and Veterinary Care

When inducing an orthopedic disease in dogs, one must first evaluate the dogs to be certain that natural bone and joint diseases are absent. Radiography is used to diagnose hip dysplasia, osteochondrosis, osteoarthritis, elbow dysplasia, and patellar luxation. These are considered heritable disorders because offspring of affected parents often have them and they occur in siblings.

Ideally, the surgical suite, the radiographic diagnostic facility, and an anesthesia recovery box lined with foam-rubber padding should be located near the primary housing facility. The floors of both the orthopedic research facility and the primary housing should be kept dry and have a nonslippery surface to provide good, steady footing.

The amount of food consumed should be monitored because excess body weight will exacerbate orthopedic conditions. Limiting food consumption during the growth period has been shown to reduce signs of orthopedic disease in dogs that mature at greater than 30 lb (Kealy et al., 1992). Dogs that refuse to eat because of pain might require a palatable high-energy food to maintain body weight. Human socialization is desirable to allow caregivers to detect abnormalities more readily and to facilitate handling and, when necessary, treatment.

Mild exercise, such as walking, is beneficial to keep muscles limber, promote bone formation, and increase lubrication and nutrition of joints. However, excessive exercise aggravates pain and causes further bone or joint damage. Anti-inflammatory drugs, given with food, can be used to relieve pain. Glucocorticosteroids, although potent anti-inflammatory agents that relieve pain, can also accelerate disease progression and should be used only in advanced cases of joint disease. Warm packs can ease the pain of chronic osteoarthritis. Dogs affected with skeletal diseases should be kept warm and dry, although pain associated with a recent injury can be eased by applying crushed ice in a plastic bag to the affected region.

Reproduction

Dogs with joint and bone diseases can generally be bred, although it might be necessary to guide and hold a male affected with moderate or severe hip dysplasia. If the orthopedic problem is so severe that mating is not possible, artificial insemination can be used.

Radiation Injury

Clinical Features

Radiation is commonly used in experimental protocols involving dogs. Total-body irradiation (TBI) is generally delivered by a cobalt-60 source or medical x-ray therapy machine. Doses of radiation up to 2 Gy can result in signs of illness related to mild gastrointestinal toxicity and decreased white-cell counts. At doses of 2-4 Gy, signs become progressively more severe. Doses greater than 4 Gy cause destruction of bone marrow, loss of circulating blood cells, immunosuppression, increased tendency to bleed, and moderate to severe gastrointestinal toxicity. Bone-marrow transplantation can prevent severe clinical signs and death in dogs. The high radiation doses are similar to the doses that human transplantation patients receive.

Several side effects occur in dogs that survive for long periods after TBI and bone-marrow rescue (Ladiges et al., 1989): pancreatic fibrosis, malabsorption and malnutrition, radiation-induced cataracts, and malignancies. A consistent finding is graying of the hair.

Radionuclides that are ingested, inhaled, or injected rarely produce signs of illness. However, knowledge of the chemical form and metabolism of the radionuclide is necessary to determine possible side effects. For example, inhaled particles of oxides of cesium-144 are relatively insoluble in the lungs and potentially remain there for some time. Signs of radiation pneumonitis might then be expected (Mauderly et al., 1980). Conversely, strontium-90 as a chloride is relatively soluble in the lungs. When inhaled, it is translocated to the bones, where it can cause prolonged thrombocytopenia and neutropenia (Gillett et al., 1987).

Types of radiation. Radiation emissions can be alpha particles, beta particles, gamma rays, and x rays. The distinctions between those emissions are important for providing care for laboratory animals.

Alpha emissions from radionuclides, such as plutonium or americium, are generally high-energy emissions, but they travel very short distances in tissue. These radionuclides are rarely used in animals unless the study is specifically intended to assess metabolic or biologic effects of alpha emissions. No special precautions are needed for direct contact with animals contaminated with alpha-particle-emitting radionuclides because the radiation energy is absorbed within the animals' tissues. However, personnel should wear disposable clothing, shoe covers, gloves, eye protection, and respiratory protection to prevent inadvertent ingestion of, inhalation of, or wound contamination with alpha particles from contaminated feces, urine, bedding, cleaning water, or surfaces.

Beta-emitting radionuclides, such as cesium-144 and strontium-90, penetrate farther into animal tissues than alpha particles but still only a short distance. The same precautions should be taken as are taken for alpha particles. Dogs can usually be handled without taking further precautions 10-12 days after administration of radionuclides.

Gamma rays and x rays from internally deposited radionuclides penetrate tissues for considerable distances. These emissions can cause some radiation exposure of personnel, and it is important to know the potential exposure levels. These are generally low-energy kinds of radiation with short half-lives. Procedures for monitoring radiation must be in place to be certain that exposures of personnel are within accepted standards. The facility radiation-protection officer should participate in planning of animal-care procedures.

Disposal of radioactive wastes is regulated by both federal and state governments. It is important to have procedures in place for collecting, packaging, and labeling radioactive wastes before studies are initiated.

Biohazards associated with radioactivity. Dogs exposed to external radiation sources do not pose a hazard to personnel once exposure is complete; the concern is for the effects on the health of the exposed animals. However, dogs that are administered radionuclides by ingestion, injection, or inhalation might present a continuing hazard to personnel because the radionuclide will be excreted in feces, in urine, and in some instances in exhaled air for some period after exposure. Standard operating procedures must be developed and followed for collecting and disposing of all contaminated materials to protect animals and personnel. Animal health is of immediate concern only when large quantities of radionuclides are given.

Husbandry and Veterinary Care

Dogs exposed to external radiation can be housed in the usual manner (see Chapter 3); however, it is critical that immunosuppressed dogs be protected from other dogs that might harbor pathogens. Dogs given internally deposited radionuclides should be housed individually. To facilitate collection of contaminated excreta and cage-cleaning water, the cages should be designed for collection of urine and feces and should be easy to clean. Dog rooms must have adequate ventilation, and ventilated air should not be recirculated. It might also be necessary to filter exhaust air. To prevent cross-contamination and simplify monitoring, it is recommended that dogs exposed to the same radionuclide be housed in the same room.

Clinical observations and frequent peripheral-blood-cell counts are useful for monitoring dogs exposed to large doses of radiation. Treatment for reduced numbers of blood cells is supportive, and euthanasia should be considered if illness becomes too severe. Marrow "rescue" can prevent severe illness. Supportive care should consist of aggressive antibiotic and fluid therapy, and a semiliquid diet is necessary during the immediate post-irradiation period. Euthanasia should be considered in long-term survivors experiencing pancreatic fibrosis, malignancies, or pneumonitis.

Reproduction

Dogs that have received TBI are usually sterile. Lower doses of radiation have variable effects on reproduction.

Gene Therapy

Gene therapy can be used to correct inborn errors of metabolism, hemoglobinopathies, and blood factor A deficiencies; to insert genes into normal cells of the host (e.g., marrow stem cells) to increase their resistance to the toxic effects of chemotherapy; to introduce genes into cancer cells that will restore suppressor-gene function or neutralize the function of activated oncogenes; and to induce tolerance to transplantation antigens by transferring genes that code for such antigens (Anderson, 1984). The use of the dog as a preclinical, large, random-bred animal model has set the stage for clinical gene therapy. A number of target tissues for gene therapy have been used; this section will cover three of them.

Hematopoietic Stem Cells

In preparation for gene transfer, marrow is aspirated while the dog is under general anesthesia. The hair over the shoulder and hip joints is clipped. The skin is cleaned with povidone iodine, washed with 70 percent ethyl alcohol, and cleansed with sterile Ringer's solution. Under sterile conditions, a needle 20 cm long and 2.5 mm in internal diameter is inserted into the marrow cavity through the proximal intertubercular groove of the humerus or trochanteric fossa of the femur. The needle is then connected with polyvinyl tubing to a suction flask, and marrow is aspirated by placing a suction flask, which contains tissue-culture medium and preservative-free heparin, under negative pressure with a pump. The procedure can be completed on all four limbs in approximately 20 minutes, during which 70-80 ml of a mixture of blood and bone marrow is collected. The marrow suspension is then passed through stainless-steel screens with 0.307- and 0.201-mm mesh diameters. A 1 ml sample is taken for marrow cell counts, and the remainder of the marrow is placed in plastic containers. The aspiration procedure is well tolerated without any sequelae. Dogs are capable of walking unimpaired after recovery from anesthesia.

Nucleated marrow cells are then cocultivated with virus-producing packaging cells at a ratio of 2:1 for 24 hours in 850-ml roller bottles. The gene-containing vector is replication-defective. Retrovirus-producing packaging cells are seeded in roller bottles 48 hours before the addition of marrow and are cultured in vitro with established techniques. After cocultivation, marrow cells are used to boost long-term cultures established 1 week earlier. The cultures are harvested after 6 days of incubation, and marrow cells are carefully removed without dislodging the virus-producing packaging cells, washed, resuspended in serum-free medium, and infused intravenously into the dog from which the marrow was taken.

In preparation for the infusion, the dog is exposed to total-body irradiation to create room for the infused marrow to seed. Total-body irradiation is administered at doses of 4-10 Gy and is usually delivered at a rate of 7 cGy/minute from two opposing cobalt-60 sources. For that purpose, an unanesthetized dog is housed in a polyurethane cage that is midway between the two cobalt-60 sources. The long axis of the cage is perpendicular to a line between the sources. After irradiation, the dog is returned to the animal-care facility for supportive care. Total-body irradiation can cause nausea, vomiting, and diarrhea. Its destruction of normal marrow leads to a disappearance of red cells, white cells, and platelets. The temporary absence of those blood components produces a risk of anemia, infection, and bleeding that persists unless the dog receives a marrow graft and the graft begins to function. Dogs are monitored daily and receive parenteral fluids and electrolytes as required. Appropriate preoperative and postoperative antibiotics are routinely used to prevent and treat infections. Platelet and red-cell transfusions are given as needed. Marrow-graft function is monitored by evaluating daily blood counts.

The success of gene transfer can be assessed by repeated aspiration of marrow under general anesthesia and examination of the samples for the appropriate marker gene with culture techniques, the polymerase chain reaction, or other appropriate methods (Stead et al., 1988). Peripheral blood cells can be tested in a similar manner, as can lymph node lymphocytes and pulmonary macrophages (Stead et al., 1988).

Skin Keratinocytes

Skin keratinocytes provide another good target for gene insertion. For some gene products, such as adenosine deaminase, gene transfer can take place in any replicating tissue. A 2×1.5-cm skin biopsy is obtained from the recipient under general anesthesia. Keratinocytes are derived from the biopsy material and cocultivated in vitro with replication-deficient retroviral vectors that contain the gene of interest. Keratinocytes are then cultured in a liquid-air interface culture, which gives rise to the various layers of skin in an in vitro system. After some time in culture, the skin grown in vitro is transplanted into a prepared bed on the flank of the dog under general anesthesia. The transplant site is treated with topical antibiotic powder, protected by nonadhering dressing, and inspected daily by the investigators. Generally, the skin grows in and is functional in 3-4 weeks. Punch biopsies of 2-3 mm allow assessment of gene transfer (Flowers et al., 1990).

Smooth Muscle Transplantation

Because of their location, genetically modified vascular smooth muscle cells can be particularly useful for the treatment of some diseases (e.g., hemophilia). Studies have demonstrated that vascular smooth muscle cells are easily obtained, cultured, and genetically modified and replaced and provide a good target tissue for gene therapy that involves both secreted and nonsecreted proteins (Lim et al., 1991). A segment of femoral artery or vein is surgically removed from a dog for preparation of smooth muscle cell cultures. The procedure of removing femoral artery and vein segments will not compromise the dog, because there is extensive collateral circulation in this region. With the dog under general anesthesia, as long a segment of vessel as possible (at least 2 cm) is isolated from the circulation with ligatures. Any side branches in the two ends are permanently ligated before the vessel is removed. The smooth muscle cells are isolated, cultured, and infected with replication-defective amphotropic retroviruses that carry the genes of interest, in accordance with National Institutes of Health recombinant-DNA guidelines. The genetically modified smooth muscle cells are returned to the animal from which they were obtained. With the dog once again under general anesthesia, the transduced cells are seeded into the left and right carotid arteries and into the remaining femoral arteries (Lim et al., 1991).

REFERENCES

Ackerman, N., R. Burk, A. W. Hahn, and H. M. Hayes, Jr. 1978. Patent ductus arteriosus in the dog: A retrospective study of radiographic, epidemiologic, and clinical findings. Am. J. Vet. Res. 39:1805-1810.

Andersen, A. C., and M. E. Simpson. 1973. The Ovary and Reproductive Cycle of the Dog (Beagle). Los Altos, Calif.: Geron-X, Inc. 290 pp.

Anderson, W. F. 1984. Prospects for human gene therapy. Science 226:401-409.

Arbulu, A., S. N. Ganguly, and E. Robin. 1975. Tricuspid valvulectomy without prosthetic replacement: Five years later. Surg. Forum 26:244-245.

AVMA (American Veterinary Medical Association). 1993. 1993 Report of the AVMA Panel on Euthanasia. J. Am. Vet. Med. Assoc. 202:229-249.

Beierwaltes, W. H., and R. H. Nishiyama. 1968. Dog thyroiditis: Occurrence and similarity to Hashimoto's struma. Endocrinology 83:501-508.

Bell, S. C., S. D. Carter, and D. Bennet. 1991. Canine distemper viral antigens and antibodies in dogs with rheumatoid arthritis. Res. Vet. Sci. 50:64-68.

Ben, L. K., J. Maselli, L. C. Keil, and I. A. Reid. 1984. Role of the renin-angiotensin system in the control of vasopressin and ACTH secretion during the development of renal hypertension in dogs. Hypertension 6:35-41.

Bice, D. E., and B. A. Muggenburg. 1985. Effect of age on antibody responses after lung immunization. Am. Rev. Respir. Dis. 132:661-665.

Blum, J. R., L. C. Cork, J. M. Morris, J. L. Olson, and J. A. Winkelstein. 1985. The clinical manifestations of a genetically determined deficiency of the third component of complement in the dog. Clin. Immunol. Immunopathol. 34:304-315.

Bonagura, J. D., ed. 1986. Section 4: Cardiovascular diseases. Pp. 319-424 in Current Veterinary Therapy. IX. Small Animal Practice, R. W. Kirk, ed. Philadelphia: W. B. Saunders.

Bovée, K. C., M. P. Littman, F. Saleh, R. Beeuwkes, W. Mann, P. Koster, and L. B. Kinter. 1986. Essential hereditary hypertension in dogs: A new animal model. J. Hypertens. 4(Suppl. 5):S172-S173.

Brooks, D. P., and T. A. Fredrickson. 1992. Use of ameroid constrictors in the development of renin-dependent hypertension in dogs. Lab. Anim. Sci. 42:67-69.

Brooks, D. P., T. A. Fredrickson, P. F. Koster, and R. R. Ruffolo, Jr. 1991. Effect of the dopamine β-hydroxylase inhibitor, SK&F 102698, on blood pressure in the 1-kidney, 1-clip hypertensive dog. Pharmacology 43:90-95.

Brown-Séquard, E. 1856. Recherches expérimentales sur la physiologie et la pathologie des capsules surrénales. Arch. Gén. Méd. (Sér. 5)8(II):385-401.

Buchanan, J. W. 1992. Causes and prevalence of cardiovascular disease. Pp. 647-654 in Current Veterinary Therapy XI, R. W. Kirk and J. D. Bonagura, eds. Philadelphia: W. B. Saunders.

Bull, R. W., R. Schirmer, and A. J. Bowdler. 1971. Autoimmune hemolytic disease in the dog. J. Am. Vet. Med. Assoc. 159:880-884.

Campbell, K. L. 1991. Immunoglobulin A deficiency in the dog: A retrospective study of 155 cases (1983-1990). Canine Pract. 16(4):7-11.

Capen, C. C., and S. L. Martin. 1989. The thyroid gland. Pp. 58-91 in Veterinary Endocrinology and Reproduction, 4th ed., L. E. McDonald and M. H. Pineda, eds. Philadelphia: Lea & Febiger.

Carlson, G. P. 1989. Fluid, electrolyte, and acid-base balance. Pp. 543-575 in Clinical Biochemistry of Domestic Animals, 4th ed., J. J. Kaneko, ed. San Diego: Academic Press.

Carter, S. D., S. C. Bell, A. S. M. Bari, and D. Bennett. 1989. Immune complexes and rheumatoid factors in canine arthritides. Ann. Rheum. Dis. 48:986-991.

Chester, D. K. 1987. The thyroid gland and thyroid diseases. Pp. 83-120 in Small Animal Endocrinology, F. H. Drazner, ed. New York: Churchill Livingstone.

Chick, T. W., S. E. Goldblum, N. D. Smith, C. Butler, B. J. Skipper, J. A. Winkelstein, L. C. Cork, and W. P. Reed. 1984. Pneumococcal-induced pulmonary leukostasis and hemodynamic changes: Role of complement and granulocytes. J. Lab. Clin. Med. 103:180-192.

Cork, L. C., J. M. Morris, J. L. Olson, S. Krakowka, A. J. Swift, and J. A. Winkelstein. 1991. Membranoproliferative glomerulonephritis in dogs with a genetically determined deficiency of the third component of complement. Clin. Immunol. Immunopathol. 60:455-470.

Cummings, J. F., ed. 1979. Part XIII: Nervous system. Pp. 107-178 in Spontaneous Animal Models of Human Disease, vol. II, E. J. Andrews, B. C. Ward, and N. H. Altman, eds. New York: Academic Press.

DePaolo, L. V., and E. J. Masoro. 1989. Endocrine hormones in laboratory animals. Pp. 279-308 in The Clinical Chemistry of Laboratory Animals, W. F. Loeb and F. W. Quimby, eds. New York: Pergamon Press.

De Reeder, E. G., N. Girard, R. E. Poelmann, J. C. Van Munsteren, D. F. Patterson, and A. C. Gittenberger-de Groot. 1988. Hyaluronic acid accumulation and endothelial cell detachment in intimal thickening of the vessel wall: The normal and genetically defective ductus arteriosus. Am. J. Pathol. 132:574-585.

De Rick, A., F. M. Belpaire, M. G. Bogaert, and D. Mattheeuws. 1978. Plasma concentrations of digoxin and digitoxin during digitalization of healthy dogs and dogs with cardiac failure. Am. J. Vet. Res. 39:811-815.

DiBartola, S. P., M. J. Tarr, D. M. Webb, and U. Giger. 1990. Familial renal amyloidosis in Chinese Shar Pei dogs. J. Am. Vet. Med. Assoc. 197:483-487.

Dodds, W. J. 1983. Immune-mediated diseases of the blood. Adv. Vet. Sci. Comp. Med. 27:163-196.

Dodds, W. J. 1988. Third international registry of animal models of thrombosis and hemorrhagic diseases. ILAR News 30:R1-R32.

Dodds, W. J. 1989. Hemostasis. Pp. 274-315 in Clinical Biochemistry of Domestic Animals, 4th ed., J. J. Kaneko, ed. San Diego: Academic Press.

Dodds, W. J. 1992. Bleeding disorders. Pp. 765-777 in Handbook of Small Animal Practice, 2d ed., R. V. Morgan, ed. New York: Churchill Livingstone.

Dougherty, S. H. 1986. Implant infections. Pp. 276-289 in Handbook of Biomaterials Evaluation, A. F. von Recum, ed. New York: Macmillan.

Drazner, F. H. 1987a. The adrenal cortex. Pp. 201-277 in Small Animal Endocrinology, F. H. Drazner, ed. New York: Churchill Livingstone.

Drazner, F. H., ed. 1987b. Small Animal Endocrinology. New York: Churchill Livingston. 508 pp.

Eigenmann, J. E. 1985. Acromegaly. Model no. 311 in A Handbook: Animal Models of Human Disease, fascicle 14, C. C. Capen, T. C. Jones, and G. Migaki, eds. Washington, D.C.: Registry of Comparative Pathology, Armed Forces Institute of Pathology.

Eigenmann, J. E. 1989. Pituitary-hypothalamic diseases. Pp. 1579-1609 in Textbook of Veterinary Internal Medicine, vol. 2, 3rd ed., S. J. Ettinger, ed. Philadelphia: W. B. Saunders.

Ettinger, S. J., ed. 1989. Textbook of Veterinary Internal Medicine, vol. 2, 3rd. ed. Philadelphia: W.B. Saunders. 1,237 pp.

Eyster, G. E. 1992. Congenital diseases. Pp. 63-69 in Handbook of Small Animal Practice, 2d ed., R. V. Morgan, ed. New York: Churchill Livingstone.

Feldman, E. C. 1989. Adrenal gland disease. Pp. 1721-1774 in Textbook of Veterinary Internal Medicine, vol. 2, 3rd ed., S. J. Ettinger, ed. Philadelphia: W. B. Saunders.

Feldman, E. C., and R. W. Nelson. 1987. Canine and Feline Endocrinology and Reproduction. Philadelphia: W. B. Saunders. 564 pp.

Felsburg, P. J., L. T. Glickman, and P. F. Jezyk. 1985. Selective IgA deficiency in the dog. Clin. Immunol. Immunopathol. 36:297-305.

Ferguson, D. C. 1986. Thyroid hormone replacement therapy. Pp. 1018-1019 in Current Veterinary Therapy IX, R. W. Kirk, ed. Philadelphia: W. B. Saunders.

Ferrario, C. M., C. Blumle, G. R. Nadzam, and J. W. McCubbin. 1971. An externally adjustable renal artery clamp. J. Appl. Physiol. 31:635-637.

Fischer, C. A. 1989. Geriatric ophthalmology. Vet. Clinics N. Am. 19(1):103-123.

Fixler, D. E., J. P. Archie, D. J. Ullyot, G. D. Buckberg, and J. I. E. Hoffman. 1973. Effects of acute right ventricular systolic hypertension on regional myocardial blood flow in anesthetized dogs. Am. Heart J. 85:491-500.

Flowers, M. E. D., M. A. R. Stockschlaeder, F. G. Schuening, D. Niederwieser, R. Hackman, A. D. Miller, and R. Storb. 1990. Long-term transplantation of canine keratinocytes made resistant to G418 through retrovirus-mediated gene transfer. Proc. Natl. Acad. Sci. USA 87:2349-2353.

Fraser, C. M., J. A. Bergeron, A. Mays, and S. E. Aiello, eds. 1991. Heart disease. Pp. 40-52 in The Merck Veterinary Manual: A Handbook of Diagnosis, Therapy, and Disease Prevention for the Veterinarian, 7th ed. Rahway, N.J.: Merck & Co.

Gardner, T. J., and D. L. Johnson. 1988. Cardiovascular system. Pp. 74-113 in Experimental Surgery and Physiology: Induced Animal Models of Human Disease, M. M. Swindle and R. J. Adams, eds. Baltimore: Williams & Wilkins.

Gelatt, K. N., ed. 1991. Veterinary Ophthalmology, 2d ed. Philadelphia: Lea & Febiger. 765 pp.

Gillett, N. A., B. A. Muggenburg, B. B. Boecker, F. F. Hahn, F. A. Seiler, A. H. Rebar, R. K. Jones, and R. O. McClellan. 1987. Single inhalation exposure to $^{90}SrCl_2$ in the beagle dog: Hematological effects. Radiat. Res. 110:267-288.

Gittenberger-de Groot, M. D., J. L. M. Strengers, M. Mentink, R. E. Poelmann, and D. F. Patterson. 1985. Histologic studies on normal and persistent ductus arteriosus in the dog. J. Am. Coll. Cardiol. 6:394-404.

Goldston, R. T., ed. 1989. Geriatrics and gerontology. Vet. Clin. N. Am. 19(1):1-202.

Gosselin, S. J., C. C. Capen, S. L. Martin, and S. Krakowka. 1982. Autoimmune lymphocytic thyroiditis in dogs. Vet. Immunol. Immunopathol. 3:185-201.

Grindem, C. B., and K. H. Johnson. 1983. Systemic lupus erythematosus: Literature review and report of 42 new canine cases. J. Am. Anim. Hosp. Assoc. 19:489-503.

Guyton, A. C. 1991. Dominant role of the kidneys in long-term regulation of arterial pressure and in hypertension: The integrated system for pressure control. Pp. 205-220 in Textbook of Medical Physiology, 8th ed. Philadelphia: W. B. Saunders.

Haley, P. J., F. F. Hahn, B. A. Muggenburg, and W. C. Griffith. 1989. Thyroid neoplasms in a colony of beagle dogs. Vet. Pathol. 26:438-441.

Hall, R. L., and U. Giger. 1992. Disorders of red blood cells. Pp. 715-733 in Handbook of Small Animal Practice, 2d ed., R. V. Morgan, ed. New York: Churchill Livingstone.

Harvey, J. W. 1989. Erythrocyte metabolism. Pp. 186-234 in Clinical Biochemistry of Domestic Animals, 4th ed., J. J. Kaneko, ed. San Diego: Academic Press.

Haskins, M. E., R. J. Desnick, N. DiFerrante, P. F. Jezyk, and D. F. Patterson. 1984. ß-glucuronidase deficiency in a dog: A model of human mucopolysaccharidosis VII. Pediatr. Res. 18:980-984.

Healy, P. J., B. R. H. Farrow, F. W. Nicholas, K. Hedberg, and R. Ratcliffe. 1984. Canine fucosidosis: A biochemical and genetic investigation. Res. Vet. Sci. 36:354-359.

Hegreberg, G. A., G. A. Padgett, J. R. Gorham, and J. B. Henson. 1969. A connective tissue disease of dogs and mink resembling the Ehlers-Danlos syndrome of man. II. Mode of inheritance. J. Hered. 60:249-254.

Hegreberg, G. A., G. A. Padgett, R. L. Ott, and J. B. Henson. 1970. A heritable connective tissue disease of dogs and mink resembling the Ehlers-Danlos syndrome of man. I. Skin tensile strength properties. J. Invest. Dermatol. 54:377-380.

Helper, L. C. 1989. Magrane's Canine Ophthalmology, 4th ed. Philadelphia: Lea & Febiger. 297 pp.

Hsu, W. H., and M. H. Crump. 1989. The adrenal gland. Pp. 202-230 in Veterinary Endocrinology and Reproduction, 4th ed., L. E. McDonald and M. H. Pineda, eds. Philadelphia: Lea & Febiger.

Järvinen, A.-K. 1981. Urogenital tract infection in the bitch. Vet. Res. Commun. 4:253-269.

Jezyk, P. F., P. J. Felsburg, M. E. Haskins, and D. F. Patterson. 1989. X-linked severe combined immunodeficiency in the dog. Clin. Immunol. Immunopathol. 52:173-189.

Johnson, J. P., R. H. McLean, L. C. Cork, and J. A. Winkelstein. 1986. Animal model: Genetic analysis of an inherited deficiency of the third component of complement in Brittany spaniel dogs. Am. J. Med. Genet. 25:557-562.

Kaneko, J. J. 1987. Critical review. Animal models of inherited hematologic disease. Clin. Chim. Acta 165:1-19.

Kaneko, J. J. 1989. Carbohydrate metabolism and its diseases. Pp. 44-85 in Clinical Biochemistry of Domestic Animals, 4th ed., J. J. Kaneko, ed. San Diego: Academic Press.

Kaplan, A. V., and F. W. Quimby. 1983. A radiolabeled staphylococcal protein A assay for detection of anti-erythrocyte IgG in warm agglutinin autoimmune hemolytic anemia of dogs and man. Vet. Immunol. Immunopathol. 4:307-317.

Kaswan, R. L., C. L. Martin, and D. L. Dawe. 1985. Keratoconjunctivitis sicca: Immunological evaluation of 62 canine cases. Am. J. Vet. Res. 46: 376-383.

Kealy, R. D., S. E. Olsson, K L. Monti, D. F. Lawler, D. N. Biery, R. W. Helms, G. Lust, and G. K. Smith. 1992. Effects of limited food consumption on the incidence of hipdysplasia in growing dogs. J. Am. Vet. Med. Assoc. 201:857-863.

Kesel, M. L., and D. H. Neil. 1990. Restraint and handling of animals. Pp. 1-30 in Clinical Textbook for Veterinary Technicians, 2d ed., D. M. McCurnin, ed. Philadelphia: W. B. Saunders.

Kirk, R. W., and S. I. Bistner. 1985. Metabolic emergencies. Pp. 138-149 in Handbook of Veterinary Procedures and Emergency Treatment, 4th ed. Philadelphia: W. B. Saunders.

Klag, A. R., U. Giger, and F. S. Shofer. 1993. Idiopathic immune-mediated hemolytic anemia in dogs: 42 cases (1986-1990). J. Am. Vet. Med. Assoc. 202:783-788.

Knight, D. H., D. F. Patterson, and J. Melbin. 1973. Constriction of the fetal ductus arteriosus induced by oxygen, acetylcholine, and norepinephrine in normal dogs and those genetically predisposed to persistent patency. Circulation 47:127-132.

Knoll, J. S. 1992. Disorders of white blood cells. Pp. 735-749 in Handbook of Small Animal Practice, 2d ed., R. V. Morgan, ed. New York: Churchill Livingstone.

Krakowka, S., L. C. Cork, J. A. Winklestein, and M. K. Axthelm. 1987. Establishment of central nervous system infection by canine distemper virus: Breach of the blood-brain barrier and facilitation by antiviral antibody. Vet. Immunol. Immunopathol. 17:471-482.

Kramer, J. W. 1981. Inherited early-onset, insulin-requiring diabetes mellitus in keeshond dogs. Am. J. Pathol. 105:194-196.

Ladiges, W. C., H. J. Deeg, J. A. Aprile, R. F. Raff, F. Schuening, and R. Storb. 1988. Differentiation and function of lymphohemopoietic cells in the dog. Pp. 307-335 in Differentiation Antigens in Lymphohemopoietic Tissues, M. Miyasaka and Z. Trnka, eds. New York: Marcel Dekker.

Ladiges, W. C., R. Storb, T. Graham, and E. D. Thomas. 1989. Experimental techniques used to study the immune system of dogs and other large animals. Pp. 103-133 in Methods of Animal Experimentation, vol. VII, part C, W. I. Gay and J. E. Heavner, eds. New York: Academic Press.

Ladiges, W. C., R. Storb, and E. D. Thomas. 1990. Canine models of bone marrow transplantation. Lab. Anim. Sci. 40:11-15.

Lage, A. L., N. A. Gillett, R. F. Gerlach, and E. N. Allred. 1989. The prevalence and distribution of proliferative and metaplastic changes in normal appearing canine bladders. J. Urol. 141:993-997.

Lange, J., B. Brockway, and S. Azar. 1991. Telemetric monitoring of laboratory animals: An advanced technique that has come of age. Lab Anim. 20(7):28-33.

Lim, C. S., G. D. Chapman, R. S. Gammon, J. B. Muhlestein, R. P. Bauman, R. S. Stack, and J. L. Swain. 1991. Direct in vivo gene transfer into the coronary and peripheral vasculatures of the intact dog. Circulation 83:2007-2011.

Lipowitz, A. J., D. D. Caywood, C. D. Newton, and M. E. Finch. 1993. Small Animal Orthopedics Illustrated: Surgical Approaches and Procedures. St. Louis: Mosby. 336 pp.

Lowseth, L. A., N. A. Gillett, R. F. Gerlach, and B. A. Muggenburg. 1990a. The effects of aging on hematology and serum chemistry values in the beagle dog. Vet. Clin. Pathol. 19(1):13-19.

Lowseth, L. A., R. F. Gerlach, N. A. Gillett, and B. A. Muggenburg. 1990b. Age-related changes in the prostate and testes of the beagle dog. Vet. Pathol. 27:347-353.

Lund, J. E., G. A. Padgett, and R. L. Ott. 1967. Cyclic neutropenia in grey collie dogs. Blood 29:452-461.

MacVean, D. W., A. W. Monlux, P. S. Anderson, Jr., S. L. Silberg, and J. F. Rozel. 1978. Frequency of canine and feline tumors in a defined population. Vet. Pathol. 15:700-715.

Maggio-Price, L., C. L. Emerson, T. R. Hinds, F. F. Vincenzi, and W. R. Hammond. 1988. Hereditary nonspherocytic hemolytic anemia in beagles. Am. J. Vet. Res. 49:1020-1025.

Mann, W. A., M. S. Landi, E. Horner, P. Woodward, S. Campbell, and L. B. Kinter. 1987. A simple procedure for direct blood pressure measurements in conscious dogs. Lab. Anim. Sci. 37:105-108.

Mauderly, J. L., and F. F. Hahn. 1982. The effects of age on lung function and structure of adult animals. Adv. Vet. Sci. Comp. Med. 26:35-77.

Mauderly, J. L., B. A. Muggenburg, F. F. Hahn, and B. B. Boecker. 1980. The effects of inhaled [144]Ce on cardiopulmonary function and histopathology of the dog. Radiat. Res. 84:307-324.

McCarthy, C. R., and J. G. Miller. 1990. OPRR Reports, May 21, 1990. Available from Office for Protection from Research Risks (OPRR), Building 31, Room 5B59, National Institutes of Health, Bethesda, MD 20892.

McDonald, L. E., and M. H. Pineda, eds. 1989. Veterinary Endocrinology and Reproduction, 4th ed. 571 pp.

Meuten, D. J., C. C. Capen, G. J. Kociba, and B. J. Cooper. 1982. Hypercalcemia of malignancy. Hypercalcemia associated with an adenocarcinoma of the apocrine glands of the anal sac. Am. J. Pathol. 108:366-370.

Meuten, D. J., C. C. Capen, and G. J. Kociba. 1986. Hypercalcemia of malignancy. Supplemental update, 1986: Model no. 143 in A Handbook: Animal Models of Human Disease, fascicle 15, C. C. Capen, T. C. Jones, and G. Migaki, eds. Washington, D.C.: Registry of Comparative Pathology, Armed Forces Institute of Pathology.

Mill, A. B., and K. L. Campbell. 1992. Concurrent hypothyroidism, IgM deficiency, impaired T-cell mitogen response, and multifocal cutaneous squamous papillomas in a dog. Canine Pract. 17(2):15-21.

Milne, K. L., and H. M. Hayes, Jr. 1981. Epidemiologic features of canine hypothyroidism. Cornell Vet. 73:3-14.

Minor, R. R., J. A. M. Wootton, D. J. Prockop, and D. F. Patterson. 1987. Genetic diseases of connective tissues in animals. Curr. Probl. Dermatol. 17:199-215.

Mizejewski, G. J., J. Baron, and G. Poissant. 1971. Immunologic investigations of naturally occurring canine thyroiditis. J. Immunol. 107:1152-1160.

Monier, J. C., C. Fournel, M. Lapras, M. Dardenne, T. Randle, and C.M. Fontaine. 1988. Systemic lupus erythematosus in a colony of dogs. Am. J. Vet. Res. 49:46-51.

Mordes, J. P., and A. A. Rossini. 1985. Animal models of diabetes mellitus. Pp. 110-137 in Joslin's Diabetes Mellitus, 12th ed., A. Marble, L. P. Krall, R. F. Bradley, A. R. Christlieb, and J. S. Soeldner, eds. Philadelphia: Lea & Febiger.

Morgan, R. V, ed. 1992. Handbook of Small Animal Practice, 2d ed. New York: Churchill Livingstone. 1,513 pp.

Moroff, S. D., A. I. Hurvitz, M. E. Peterson, L. Saunders, and K. E. Noone. 1986. IgA deficiency in Shar-Pei dogs. Vet. Immunol. Immunopathol. 13:181-188.

Nakano, K., M. M. Swindle, F. G. Spinale, K. Ishihara, S. Kanazawa, A. Smith, R. W. W. Biederman, L. Clamp, Y. Hamada, M. R. Zile, and B. A. Carabello. 1991. Depressed contractile function due to canine mitral regurgitation improves after correction of the volume overload. J. Clin. Invest. 87:2077-2086.

Nelson, A. A., and G. Woodard. 1949. Severe adrenal cortical atrophy (cytotoxic) and hepatic damage produced in dogs by feeding 2,2-bis(parachlorophenyl)-1,1-dichloroethane (DDD or TDE). Arch. Pathol. 48:387-394.

Nelson, R. W. 1989. Disorders of the endocrine pancreas. Pp. 1676-1720 in Textbook of Veterinary Internal Medicine, vol. 2, 3rd ed., S. J. Ettinger, ed. Philadelphia: W. B. Saunders.

Newton, C. D., and D. M. Nunamaker. 1985. Textbook of Small Animal Orthopaedics. Philadelphia: J. B. Lippincott. 1,140 pp.

Nichols, R., and M. E. Peterson. 1992. Hypoadenocorticism. Pp. 531-534 in Handbook of Small Animal Practice, 2d ed., R. V. Morgan , ed. New York: Churchill Livingstone.

NRC (National Research Council), Institute of Laboratory Animal Resources, Committee on Care and Use of Laboratory Animals. 1985. Guide for the Care and Use of Laboratory Animals. NIH Pub. No. 86-23. Washington, D.C.: U.S. Department of Health and Human Services. 83 pp.

NRC (National Research Council), Institute of Laboratory Animal Resources, Committee on Immunologically Compromised Rodents. 1989. Introduction. Pp. 1-35 in Immunodeficient Rodents: A Guide to Their Immunobiology, Husbandry, and Use. Washington, D.C.: National Academy Press.

Ogilive, G. K., W. M. Haschek, S. J. Withrow, R. C. Richardson, H. J. Harvey, R. A. Henderson, J. D. Fowler, A. M. Norris, J. Tomlinson, D. McCaw, J. S. Klausner, R. W. Reschke, and B. C. McKiernan. 1989. Classification of primary lung tumors in dogs: 210 cases (1975-1985). J. Am. Vet. Med. Assoc. 195:106-108.

O'Kane, H. O., A. S. Geha, R. E. Kleiger, T. Abe, M. T. Salaymeh, and A. B. Malik. 1973. Stable left ventricular hypertrophy in the dog. Experimental production, time course, and natural history. J. Thorac. Cardiovasc. Surg. 65:264-271.

Oliver, J. E. Jr., and M. D. Lorenz. 1993. Appendix. Pp. 374-393 in Handbook of Veterinary Neurology, 2d ed. Philadelphia: W. B. Saunders.

O'Neil, K. M., H. D. Ochs, S. R. Heller, L. C. Cork, J. M. Morris, and J. A. Winkelstein. 1988. Role of C3 in humoral immunity. Defective antibody production in C3-deficient dogs. J. Immunol. 140:1939-1945.

Patterson, D. F. 1968. Epidemiologic and genetic studies of congenital heart disease in the dog. Circ. Res. 23:171-202.

Patterson, D. F. 1984. Two hereditary forms of ventricular outflow obstruction in the dog: Pulmonary valve dysplasia and discrete subaortic stenosis. Pp. 43-63 in Congenital Heart Disease: Causes and Processes, J. J. Nora and A. Takao, eds. Mt. Kisco, N.Y.: Future Publishing Co.

Patterson, D. F., R. L. Pyle, J. W. Buchanan, E. Trautvetter, and D. A. Abt. 1971. Hereditary patent ductus arteriosus and its sequelae in the dog. Circ. Res. 29:1-13.

Patterson, D. F., R. L. Pyle, L. Van Mierop, J. Melbin, and M. Olson. 1974. Hereditary defects of the conotruncal septum in keeshond dogs: Pathologic and genetic studies. Am. J. Cardiol. 34:187-205.

Patterson, D. F., M. E. Haskins, and W. R. Schnarr. 1981. Hereditary dysplasia of the pulmonary valve in beagle dogs: Pathologic and genetic studies. Am. J. Cardiol. 47:631-641.

Patterson, D. F., M. E. Haskins, and P. F. Jezyk. 1982. Models of human genetic disease in domestic animals. Adv. Hum. Genet. 12:263-339.

Patterson, D. F., T. Pexieder, W. R. Schnarr, T. Navratil, and R. Alaili. 1993. A single major-gene defect underlying cardiac conotruncal malformations interferes with myocardial growth during embryonic development: Studies in the CTD line of keeshond dogs. Am. J. Hum. Genet. 52:388-397.

Petersen, J. C., R. R. Linartz, R. L. Hamlin, and R. E. Stoll. 1988. Noninvasive measurement of systemic arterial blood pressure in the conscious beagle dog. Fundam. Appl. Toxicol. 10:89-97.

Peterson, M. E., and D. C. Ferguson. 1989. Thyroid disease. Pp. 1632-1675 in Textbook of Veterinary Internal Medicine, vol. 2, 3rd ed., S. J. Ettinger, ed. Philadelphia: W. B. Saunders.

PHS (Public Health Service). 1986. Public Health Service Policy on Humane Care and Use of Laboratory Animals. Washington, D.C.: U.S. Department of Health and Human Services. 28 pp. Available from the Office for Protection from Research Risks, Building 31, Room 4B09, NIH, Bethesda, MD 20892.

Plechner, A. J. 1979. IgM deficiency in 2 doberman pinschers. Mod. Vet. Pract. 60:150.

Pyle, R. L., D. F. Patterson, and S. Chacko. 1976. The genetics and pathology of discrete subaortic stenosis in the Newfoundland dog. Am. Heart J. 92:324-334.

Quimby, F. W. 1981. Canine systemic lupus erythematosus. Pp. 175-184 in Immunologic Defects in Laboratory Animals, vol. 2, M. E. Gershwin and B. Merchant, eds. New York: Plenum Press.

Quimby, F. W., and R. S. Schwartz. 1978. The etiopathogenesis of systemic lupus erythematosus. Pathobiol. Annu. 8:35-59.

Quimby, F. W., C. Jensen, D. Nawrocki, and P. Scollin. 1978. Selected autoimmune diseases in the dog. Vet. Clin. N. Am. 8(4):665-682.

Quimby, F. W., R. S. Schwartz, T. Poskitt, and R. M. Lewis. 1979. A disorder of dogs resembling Sjögren's syndrome. Clin. Immunol. Immunopathol. 12:471-476.

Quimby, F. W., C. Smith, M. Brushwein, and R.W. Lewis. 1980. Efficacy of immunoserodiagnostic procedures in the recognition of canine immunologic diseases. Am. J. Vet. Res. 41:1662-1666.

Rajatanavin, R., S.-L. Fang, S. Pino, P. Laurberg, L. Braverman, M. Smith, and L. P. Bullock. 1989. Thyroid hormone antibodies and Hashimoto's thyroiditis in mongrel dogs. Endocrinology 124:2535-2540.

Renshaw, H. W., and W. C. Davis. 1979. Canine granulocytopathy syndrome. An inherited disorder of leukocyte function. Am. J. Pathol. 95:731-744.

Rivas, A. L., L. Tintle, E. S. Kimball, J. Scarlett, and F. W. Quimby. 1992. A canine febrile disorder associated with elevated interleukin-6. Clin. Immunol. Immunopathol. 64:36-45.

Ross, L. A. 1989. Hypertensive disease. Pp. 2047-2056 in Textbook of Veterinary Internal Medicine, vol. 2, 3rd ed., S. J. Ettinger, ed. Philadelphia: W. B. Saunders.

Roth, J. A., L. G. Lomax, N. Altszuler, J. Hampshire, M. I. Kaeberle, M. Shelton, D. D. Draper, and A. E. Ledet. 1980. Thymic abnormalities and growth hormone deficiency in dogs. Am. J. Vet. Res. 41:1256-1262.

Schrader, L. A. 1988. Hypoadrenocorticism. Pp. 543-546 in Handbook of Small Animal Practice, 2d ed., R. V. Morgan, eds. New York: Churchill Livingstone.

Schwartz, R. S., F. W. Quimby, and J. André-Schwartz. 1978. Canine systemic lupus erythematosus: Phenotypic expression of autoimmunity in a closed colony. Pp. 287-294 in Genetic Control of Autoimmune Disease, N. R. Rose, P. Bigazzi, and N. Warner, eds. New York: Elsevier-North Holland.

Shull, R. M., R. J. Munger, E. Spellacy, C. W. Hall, G. Constantopoulos, E. F. Neufeld. 1982. Canine α-L-iduronidase deficiency: A model of mucopolysaccharidosis I. Am. J. Pathol. 109:244-248.

Smith, G. S., and J. H. Lumsden. 1983. Review of neutrophil adherence, chemotaxis, phagocytosis and killing. Adv. Vet. Immunol. 1982 12:177-236.

Stead, R. B., W. W. Kwok, R. Storb, and A. D. Miller. 1988. Canine model for gene therapy: Inefficient gene expression in dogs reconstituted with autologous marrow infected with retroviral vectors. Blood 71:742-747.

Swindle, M. M., F. G. Spinale, A. C. Smith, R. E. Schumann, C. T. Green, K. Nakano, S. Kanasawa, K. Ishihara, M. R. Zile, and B. A. Carabello. 1991. Anesthetic and postoperative protocols for a canine model of reversible left ventricular volume overload. J. Invest. Surg. 4:339-346.

Taylor, G. N., L. Shabestari, J. Williams, C. W. Mays, W. Angus, and S. McFarland. 1976. Mammary neoplasia in a closed beagle colony. Cancer Res. 36:2740-2743.

Terman, D. S., D. Moore, J. Collins, B. Johnston, D. Person, J. Templeton, R. Poser, and F. Quimby. 1979. Detection of immune complexes in sera of dogs with rheumatic and neoplastic diseases by ^{125}I-Clq binding test. J. Comp. Pathol. 89:221-227.

Thacker, E. L., K. R. Refsal, and R. W. Bull. 1992. Prevalence of autoantibodies to thyroglobulin, thyroxine, or triiodothyronine and relationship of autoantibodies and serum concentrations of iodothyronines in dogs. Am. J. Vet. Res. 53:449-453.

Tholen, M. A., and R. F. Hoyt, Jr. 1983. Oral pathology. Pp. 39-67 in Concepts in Veterinary Dentistry. Edwardsville, Kansas: Veterinary Medicine Publishing Co.

Valentine, B. A., N. J. Winand, D. Pradhan, N. S. Moise, A. de Lahunta, J. N. Kornegay, and B. J. Cooper. 1992. Canine X-linked muscular dystrophy as an animal model of Duchenne muscular dystrophy: A review. Am. J. Med. Genetics 42:352-356.

Van Mierop, L. H. S., D. F. Patterson, and W. R. Schnarr. 1977. Hereditary conotruncal septal defects in keeshond dogs: Embryologic studies. Am. J. Cardiol. 40:936-950.

Vlahakes, G. J., K. Turley, and J. I. E. Hoffman. 1981. The pathophysiology of failure in acute right ventricular hypertension: Hemodynamic and biochemical correlations. Circulation 63:87-95.

Waye, J. W. 1960. Idiopathic thrombocytopenic purpura in a dog. Can. Vet. J. 1:569-571.

Whitney, J. C. 1967. The pathology of the canine genital tract in false pregnancy. J. Small Anim. Pract. 8:247-263.

Whittick, W. G., ed. 1990. Canine Orthopedics, 2d ed. Philadelphia: Lea & Febiger. 936 pp.

WHO (World Health Organization) Scientific Group. 1986. Primary immunodeficiency diseases. Clin. Immunol. Immunopathol. 40:166-196.

Willis, M. B. 1989. Genetics of the Dog. London: H. F. & G. Witherby. 417 pp.

Winkelstein, J. A., L. C. Cork, D. E. Griffin, J. W. Griffin, R. J. Adams, and D. L. Price. 1981. Genetically determined deficiency of the third component of complement in the dog. Science 212:1169-1170.

Winkelstein, J. A., J. P. Johnson, A. J. Swift, F. Ferry, R. Yolken, and L. C. Cork. 1982. Genetically determined deficiency of the third component of complement in the dog: *In vitro* studies on the complement system and complement-mediated serum activities. J. Immunol. 129:2598-2602.

Winkelstein, J. A., J. P. Johnson, K. M. O'Neil, and L. C. Cork. 1986. Dogs deficient in C3. Progr. Allergy 39:159-168.

Wiśniewski, H., A. B. Johnson, C. S. Raine, W. J. Kay, and R. D. Terry. 1970. Senile plaques and cerebral amyloidosis in aged dogs: A histochemical and ultrastructural study. Lab. Invest. 23:287-296.

Young, D. M., ed. 1979. Part XV: Skeletal system. Pp. 197-264 in Spontaneous Animal Models of Human Disease, vol. II, E. J. Andrews, B. C. Ward, and N. H. Altman, eds. New York: Academic Press.

Appendix

Cross Reference

Subject	Page No. in This Report	Part No. in AWRs (9 CFR)	Page No. in *Guide*
Bedding storage	26	3.1e	24
Chemicals and toxic substances	19	—	22, 25
Emergency power	18	—	46
Exercise	21-24	3.8	17
Feeding	25-26	3.9	22-23
Food storage	26	3.1e	23, 46
Handling	78	2.131	—
Housing facilities			
General construction	12-14	3.1a-b, 3.4c	42-43
Drains	14-15	3.1f	44
Lockers, washrooms, and toilet areas	13, 15	3.1g	6, 42
Physical relationship of animal facilities to laboratories	14	—	41
Power and Lighting	18	3.1d	46
Surfaces	14-15	3.1c ,3.2d	43-45
Humidity, indoor	16-17	3.2b	18-19, 45
Identification	27-28	2.38g, 2.50	27
Illumination	18	3.2c	20-21
Noise	19	—	21

Subject	Page No. in This Report	Part No. in AWRs (9 CFR)	Page No. in *Guide*
Outside runs	15-16	3.3e	—
Primary enclosures	19-20	3.6a,c,d	11-12
Procurement	52-53	2.60	34
Protocol review	76-78	2.31c,d	—
Record-keeping			27
Annual or semiannual reports	29	2.36	
Dogs on hand	28-29	2.35b-e	
Dog procurement	28-29	2.35b	
Institutional animal care and			
use committee	29	2.35a	—
Restraint	78	—	9
Sanitation	27	3.11	24-27
Social interaction	22-24	3.6c3, 3.7	12-13
Space	20-21	3.6a,c	13-17
Temperature, indoor	16-17	3.2a	18-19, 45
Training employees	2	2.32, 3.12	4–5
Transportation	29-32	3.13-3.19	—
Ventilation, indoor	17-18	3.2b	19-20, 45-46
Veterinary care			
Analgesia	66-67	2.33b	37
Anesthesia	64-66	2.33b	37
Conditioning	53	—	34-35
Emergency, weekend, and			
holiday care	29	2.33b	28
Euthanasia	70-72	2.33b	38-39
General	51	2.33, 2.4	33
Postsurgical care	69-70	2.33b	37-38
Quarantine	54-55	—	34-35
Surgery	68-69	2.31d	9-10, 37-38, 47-48
Surveillance, diagnosis, treatment,			
and control of infectious diseases	53-57	2.33b	36-37
Waste disposal	27	3.1f	26-27
Watering	26	3.10	24

Index

Acromegaly, 94, 96
Addison's disease (hyperadrenocorticism), 93, 95-96
Adenovirus, 54-57
Adipose tissue, 26
Aging, 79-81
Alaria canis, 62
Albumin, 106
Alpha-2 agonists, 67
Amebas, 58, 59
American College of Veterinary Ophthalmologists, 115
American Veterinary Medical Association (AVMA) Panel on Euthanasia, 70-71
Ameroid, 88-89
Analgesics
anti-inflammatory nonsteroidal, 67
euthanasia, 70-71
opioid, 66-67
pain alleviation, 63-68
postsurgical, 68-69
Ancylostoma spp., 44, 59-61
Anemia, 100
Anesthesia and anesthetics
alleviation of pain, 64-68
euthanasia, 70-71
gene therapy, 120, 122
general, 64-66, 69
inhalants, 64, 71
injectable drugs, 64-65, 70-71
local, 66
hypothermia, 17
neuromuscular blocking agents, 65-66
presurgical, 69

Anestrus, 41, 81
Angiotensin-converting enzyme (ACE), 87-88, 91
Animal Welfare Act, 1, 11
Animal Welfare Regulations (AWRs)
cross referenced, 131-132
defined, 1-2
Antigens, 5, 120
Antibiotics, 69
Arthritis, 5, 79
Artificial insemination, 37-39
Ascarids (*Toxocara canis; Toxascaris leonina*), 58-60
Association of American Feed Control Officials, 24, 42
Attachment formation, 44-45
Auscultation, 85
Autoimmune diseases, *see* Immunologic diseases; *and specific diseases*
Autoimmune hemolytic anemia, 102-104
Autoimmune (lymphocytic) thyroiditis, 102-104

Babesia spp., 62-63
Babesiosis, 62-63
Bacterial diseases, *see specific diseases*
Balantidium spp., 58-59
Barbiturates, 70
Bedding and resting apparatus, 11, 26-27, 30, 113
Behavior
aging, 79
aggressive, 5, 23, 30, 45-46, 67, 70
blindness and ocular pain, 115-116

fearful, 23, 45-46, 53, 67-68
herding, 5
maladaptive, 22, 53, 67
neurologic disorders, 108-109
record-keeping, 46
selection of experimental animals, 7-9
socialization, 7-9, 11, 22-24, 44-46, 53
see also Stressors and distress
Benzodiazepines, 67-68
Biohazards, 119
Bioimplants, 85
Bladder disorders, 81
Bleeding disorders, 99-101
Blindness, 114-115
Blood glucose monitoring, 94
Blood pressure, 89-90
Blood urea nitrogen (BUN), 89
Body size, 7, 25-26, 30
Body weight, 7, 25-26, 30, 39-40, 42-43, 67,
    80, 95, 99, 117
Bone marrow transplantation, 107, 117, 120-
    121
Bordetella bronchiseptica, 54-56
Breeding colonies
    deworming, 43-44
    infectious diseases, 40-41, 55-56
    neonatal care, 39-47, 56
    nutritional requirements, 42-43
    record-keeping, 46-47
    reproduction, 35-41, 83-84
    socialization, 44-46
    specific-pathogen-free, 57
    vaccination, 43-44, 55-56
    see also Specific-pathogen-free
        animals; Reproduction; Vaccines
        and vaccination
Brucella canis, 37, 55
Brucellosis, 37, 55

Cages and pens, 19-20, 30-31
Calcium derangements, 93-97
Carbon dioxide, 71
Carbon monoxide, 71
Cardiovascular diseases, 5, 61, 70, 80-87;
    see also Hypertension; *and specific
    diseases*
Catheterization, 69, 85-87, 89, 113
Cheyletiella yasguri, 58
Chemoprophylaxis, 59, 61
Code of Federal Regulations, defined, 1-2
Colostrum, 43, 104
Common variable immunodeficiency, 102-
    104
Complement (C3) deficiency, 105-106
Complete blood counts (CBCs), 86-87
Congenital heart defects, 81-84
Conotruncal defects, 82
Cornification, 39
Cryptosporidium spp., 58-59
Cyanosis, 83-84
Cyclic hematopoiesis, 98, 100

Decapitation, 71
Degenerative joint disease, 79
Demodex canis, 58
Dental diseases, 25, 79-80, 100
Deoxycorticosterone acetate (DOCA), 88
Deworming, 43-44
Diabetes mellitus, 5, 94-95
Diarrhea, 59, 61
Diazepam, 67-68
Digoxin, 80-81
Dipylidium caninum, 62
Dirofilaria immitis, 5, 59, 61
Dirofilariasis, 5, 59, 61
Distemper, 5, 54-57
Distress, *see* Stressors and distress
DLA (major histocompatiblity complex), 5
Drainage, 14-15
Drugs, *see* Analgesics; Anesthesia and
    Anesthetics; Injectable drugs; *and
    specific drugs*
Duchenne's muscular dystrophy, 110-111
Dysesthesias, 113
Dystocia, 41-42

Ear mites (*Otodectes cynotis*), 58
Echinococcus spp., 62
Echocardiography, 85, 87
Ectoparasites, 57-58; *see also* Parasitic
    diseases; *and specific ectoparasites*
Ehlers-Danlos syndrome, 12, 91-92
Ehrlichia canis, 62-63
Ehrlichiosis, 62-63
Electrocardiography, 69, 85, 87
Electrocution, 71
Embryo-transfer technology, 57
Emergency, weekend, and holiday care, 29
Endocrinologic diseases, 79, 93-97
    clinical features, 93-94
    husbandry and veterinary care, 94-97
    see also specific diseases
Endoparasites, 44, 58-63; *see also* Parasitic
    diseases; *and specific endoparasites*
Entamoeba spp., 59
Environmental controls, *see* environmental
    conditions *under* Housing
Environmental enrichment, 11, 21-24
Enzyme-linked immunosorbent assay
    (ELISA) kit, 37, 41
Erythrocyte phosphofructokinase deficiency,
    98, 100-101
Esophageal nematode (*Spirocerca lupi*), 62
Estrus, 23, 30, 35-39, 41, 81, 95-96
Euthanasia
    cardiac defects, 83, 86
    ethics, 72
    human considerations, 71-72
    inhalation methods, 71
    injection methods, 70
    lysosomal storage diseases, 108-110
    muscular dystrophy, 111-112
    necropsy examination, 40

physical methods, 71
radiation injury, 119
Exercise, *see under* Housing
Exsanguination, 71

Factor X deficiency, 98, 101
False estrus, 41
Fearful behavior, 23, 45-46, 53, 68
Filaroides spp., 58-59
Fleas, 58, 62
Food and nutrition
    aging, 79-80
    bleeding disorders, 99-100
    body size, 25-26
    body weight, 25-26, 43
    cardiovascular diseases, 85
    conditioning, 53
    contaminants, 26
    deprivation, 78
    diabetes, 95-96
    feeding programs, 25-26
    hypertension, 90-91
    labels, 24
    lysosomal storage diseases, 109-110
    muscular dystrophy, 111
    neonatal, 42-43
    neurologic diseases, 113
    nutritional content, 24-25
    organ transplantation, 107
    orthopedic diseases, 117
    pregnancy and lactation, 42
    restraint training, 78
    transportation, 29-31
Forms, 28-29
Fucosidosis, 108-109

Gases, 19, 31
Gene therapy, 79, 119-122
Genetic factors, and selection of
        experimental animals, 5-7, 79
Genetic mapping, 5-6
Geriatrics, 79-81
Geriatrics and Gerontology, 81
Giardia spp., 58-59
Glucocorticosteroids, 117
Good Laboratory Practice Standards, 2
Granulocytopathy, 102-104
Granulomatous pneumonia, 111
Growth hormones, 96, 102-104
*Guide for the Care and Use of Laboratory
        Animals (Guide),* defined, 1-2

Hair-follicle mites (*Demodex canis*), 58
Hazards, and selection of experimental
        animals, 5
Health Research Extension Act of 1985, 1
Hearing loss, 79
Heartworms (*Dirofilaria immitis*), 5, 59, 61
Helminths, 58-63
Hematologic diseases, 97-98
    bleeding disorders, 99-101
    clinical features, 6, 97-98

cyclic hematopoiesis, 100
    husbandry and veterinary care, 99-
        101
    reproduction, 101
    see also specific diseases
Hematopoietic stem cells, 120-121
Hemophilia, 97, 101, 121
Hepatitis, 54-57
Herpesvirus, 54-56
Heterodoxus spiniger, 58
Hookworms (*Ancylostoma* spp.; *Uncinaria
        stenocephala*), 44, 59-61
Housing
    aging, 80
    chemicals and toxic substances, 19
    criteria for design and construction, 12-
        14
    drainage, 14-15
    Ehlers-Danlos syndrome, 92
    environmental conditions, 16-19, 31,
        44-46, 104
    exercise, 11-12, 21-22, 83, 95, 111
    hematologic diseases, 99-101
    holding areas, 31-32
    hypertension, 91
    immunodeficiency diseases, 104-106
    indoor facilities, 14-15
    lysosomal storage diseases, 109
    muscular dystrophy, 111
    neurologic diseases, 113
    noise, 19, 39
    neurologic diseases, 113
    outdoor facilities, 14, 16
    postoperative, 85
    power and lighting, 18
    primary enclosures, 19-20, 30-31
    quarantine facilities, 54-55
    radiation injury, 119
    sheltered housing facilities, 14-16
    solitary, 22
    space recommendations, 20-21
    temperature and humidity, 16-17, 31,
        40, 80
    ventilation, 17-19, 30-31, 119
    whelping facilities, 39-40
Husbandry, *see* Bedding and resting
        apparatus; Emergency, weekend,
        and holiday care; Environmental
        enrichment; Food and nutrition;
        Housing; Identification and records;
        Record-keeping; Sanitation; Water
        and watering devices
Hyperadrenocorticism (Addison's disease),
        93, 95-96
Hypercalcemia, 93-97
Hypertension, 87-91
Hypoadrenocorticism, 93-96
Hypocalcemia, 93-96
Hypothermia, 17, 26-27
Hypothyroidism, 93
Hypoxia, 71

IATA, *see* International Air Transport
    Association
Identification and records, 11, 27-30
IgA deficiency, 102-104
IgM deficiency, 102-104
Immune thrombocytopenic purpura, 102-104
Immunologic diseases
    acquired, 102
    autoimmune, 5, 101-105
    clinical features, 6, 101-107
    complement deficiency, 105-106
    husbandry and veterinary care, 104-107
    organ transplantation, 106-107
    primary immunodeficiency, 101-105
    see also specific diseases
Immunoprophylaxis, 43
Inbreeding, 37
Induced heart defects, 84-87
Infectious diseases, 5, 53-57, 80; *see also*
    specific pathogens and specific
    diseases
Inhalant anesthetics, *see under* Anesthesia
    and anesthetics
Injectable drugs, 64-65, 70-71
Institutional animal care and use committee
    (IACUC), 29, 76-78
Instrument implantation, 70
Instruments, artificial insemination, 38-39
Insulin, 94-95
Interleukin-6 dysregulation, 102-105
International Air Transport Association
    (IATA), 2, 31
Interstate and International Certificate of
    Health Examination for Small
    Animals (USDA), 29
Intestinal fluke (*Alaria canis*), 62
Isospora spp., 58-59

Kennel cough, 54-56

Leishmania spp., 62
Leishmaniasis, 62
Leptospirosis, 55-56
Lethal injection, 70-71
Lice (*Linognathus setosus; Trichodectes*
    canis; Heterodoxus spiniger), 58
Linognathus setosus, 58
List of Licensed Dealers, 52
Lung diseases, 60, 80
Lung fluke (*Paragonimus kellicotti*), 61-62
Lysosomal storage diseases (LSDs), 6, 107-
    110; *see also specific diseases*

Major histocompatibility complex (DLA), 5
Male-female ratio, 37
Mange (*Sarcoptes scabei*), 58
Mating, *see* Reproduction
Measles, 56
Metabolic bone diseases, 25, 43
Microsatellite probes, 6
Microsporum spp., 54

Mites, 58
Mitotane, 95-96
Models, canine, 6
Monitoring, 17, 53
Motor deficits, 113
Mucopolysaccharidosis, 108-110
Muscle mass, 26
Muscular dystrophy, 110-112
Musculoskeletal diseases, 6
Mycoplasma spp., 54

Nasal mite (*Pneumonyssoides caninum*), 58
National Association of State Public Health
    Veterinarians, 55
Nematodes, 59, 61-62
Neonatal care, *see under* Breeding colonies
Neurologic diseases, 5, 6, 112-114; *see also*
    specific diseases
Neuromuscular blocking agents, 65-66
Neutropenia, 118
Nicotine, 71
Nitrous oxide, 64
Nutrition, *see* Food and nutrition

Ocular defense mechanisms, 115-116
Ocular pain, 114, 116
Oocysts, 59
Ophthalmologic diseases, 114-116; *see also*
    specific diseases
Organ transplantation, 106-107
Orthopedic diseases, 5, 116-117; *see also*
    specific diseases
Otodectes cynotis, 58
Ownership transfer, 28

Packs, 7-8, 23
Pain, 63-67, 114, 116; *see also* Analgesics;
    Anesthesia and anesthetics;
    Stressors and distress; Surgery
Paragonimus kellicotti, 61-62
Parainfluenza, 54-56
Parasitic diseases, 5, 8-9, 37, 44, 55, 57-63;
    see also specific parasites and
    specific diseases
Particulate contaminants, 31
Parturition, *see* Reproduction
Parvovirus, 5, 54-56
Patent ductus arteriosus, 81-83
Pathogens, *see* Infectious diseases; *and*
    specific pathogens
Pentatrichomonas spp., 59
Persistent truncus arteriosus, 82
Pharmacologic therapy, 87
Phenothiazines, 67-68
Pheromones, 37
Photoperiod, and reproductive cycle, 36
Physaloptera spp., 62
Physical fitness and enclosure size, 20
Physiologic monitoring and testing, 53
    fecal and blood tests for endoparasites,
    63

implantation of instruments, 70
inadequately socialized dogs, 45
induced heart defects, 85-86
pregnancy tests, 39-40
renal function in hypertensive dogs, 88-89
surgical and postsurgical, 69
Pneumonyssoides caninum, 58
Polyps, 81
Pregnancy, *see* Reproduction
Procurement, 52-53
Prostatic disorders, 81
Proteinuria, 106
Protocol review, 76-78
Protozoa, 58-63
Pseudopregnancy, 41
Public Health Service Policy on Humane Care and Use of Laboratory Animals (PHS Policy), 1-2
Pulmonary valve dysplasia, 83
Purpose-bred animals, 52, 54, 57-61, 63, 68
Pyometra, 81
Pyruvate kinase deficiency, 98, 100

Quarantine facilities, 54-55

Rabies, 9, 54-56
Radiation injury, 117-119
Radiation pneumonitis, 118
Radioactive-waste disposal, 118
Radiography, as pregnancy test, 39
Radionuclides, 118-119
Random-source animals, 52-55, 57-58, 63
Record of Disposition of Dogs and Cats (USDA), 29
Record of Dogs and Cats on Hand (USDA), 29
Record-keeping
  animal-care staff, 28
  federal regulations, 28-29
  reproduction, 38, 46-47
Reinforcement techniques, 78
Renal diseases, 80, 87-89, 91, 106
Retinal degeneration, 6, 79, 88, 90-91
Reproduction
  acromegaly, 96
  aging, 81
  anestrus, 41, 81
  artificial insemination, 37-39
  cardiovascular diseases, 83-84
  Ehlers-Danlos syndrome, 92
  estrus, 23, 30, 35-39 ,41, 81, 95-96
  false estrus, 41
  hematologic dieseases, 101
  immunologic diseases, 105-106
  lysosomal storage diseases, 110
  muscular dystrophy, 112
  natural mating, 37-38
  neurologic diseases, 114
  ophthalmologic diseases, 116

orthopedic diseases, 117
pregnancy and parturition, 39-41, 114
pseudopregnancy, 41
radiation, 119
record-keeping, 38, 46-47
reproductive cycle, 35-37, 46, 81
semen, 38-39
see also Breeding colonies
Research protocols, 76-78
Respiratory diseases, 55, 113; *see also specific diseases*
Restraint methods, 78
Restriction-fragment length polymorphisms (RFLPs), 6
Rheumatoid arthritis, 102-104
Rhipicephalus sanguineus, 57-58, 62
Ringworm (*Microsporum* spp.), 54
Roundworm (*Toxocara canis*), 44, 58-60

Sanitation, 14-15, 19, 26-27, 30, 58-63, 68-69, 119
Sarcoptes scabei, 58
Scent-marking, 37
Semen, 38-39
Sensory deficits, 113
Septicemia, 102, 107
Serum creatinine concentration, 89
Serum urea nitrogen, 106
Severe combined immunodeficiency, 102-104
Sjögren's syndrome, 102-104
Skin keratinocytes, 121
Smooth muscle transplantation, 121-122
Social contact and interaction, 7-9, 11, 22-24, 44-46, 53, 117
Sodium-to-potassium ratio, 95-96
Specific-pathogen-free (SPF) animals, 57, 59-61, 63
Spirocerca lupi, 62
Spondylosis, 79
Sterilization, surgical instruments, 68-69
Stomach nematode (*Physaloptera* spp.), 62
Stressors and distress
  allevation, 64-68
  blindness, 115
  blood-glucose response, 95
  environmental, 21-22, 53, 67-68, 95-96
  hypertension measurement, 90
  non-pain-induced, 67-68
  pain-induced, 63-64
  parturition, 41
  recognition, 63-64, 67
  signs of pain, 64
  sleep, 22
  transportation, 29-30
  treatment, 67-68
  vocalization, 44-45, 63-64
  see also Analgesics; Anesthesia and anesthetics; Behavior; Pain; Surgery
Strongyloides stercoralis, 59-60
Strychnine, 71
Subaortic stenosis, 81-83

Surgery
cardiovascular diseases, 84-87
gene therapy, 120-122
hypothermic recovering dogs, 17
pain, 63-67
postsurgical care, 69-70, 85, 91, 107
presurgical preparation, 68-69, 107
record-keeping, 28
renal failure, 89
see also Analgesics; Anesthesia and anesthetics; Pain; Stressors and distress
Systemic lupus erythematosus, 102-104

*T*aenia spp., 62
Tapeworms (*Dipylidium caninum; Echinococcus* spp.; *Taenia* spp.), 62
Tattoos, 27-28
Testicular atrophy, 81
Tetralogy of Fallot, 82
Thrombocytopenia, 118
Thrombopathia, 98, 101
Thyrogastric disease, 102-104
Thyroid atrophy, 79
Thyroiditis, 102-104
Tick (*Rhipicephalus sanguineus*), 57-58, 62
Total-body irradiation (TBI), 107, 117, 119-121
Total serum protein, 106
Toxascaris leonina, 58-60
Toxocara canis, 44, 58-60
Tracheobronchitis, 54-55
Tranquilizers, 66-68, 70; *see also specific tranquilizers*
Transplantation studies, 57
Transponders, 28
Transportation
environmental conditions, 31
health certificates, 31-32
holding areas, 31-32
food and water, 29-31
primary enclosures, 30-31
stress, 29-30
Trichodectes canis, 58
Trichomonas spp., 59
Trichuris vulpis, 59, 61
Truncus arteriosus, persistent, 82
Trypanosoma cruzi, 62

*U*ltrasonography, 39
U.S. Department of Agriculture (USDA), 11, 21, 27-28, 32, 52
U.S. Government Principles for Utilization and Care of Vertebrate Animals Used in Testing, Research, and Training, 63-64
Uncinaria stenocephala, 44, 59-61

*V*accines and vaccination
animal-care personnel, 9, 62
bleeding disorders, 101
breeding colonies, 43-44, 55-56
development, 5, 57
hematologic disorders, 101
multivalent, 55-56
schedule for pups, 56
social and behavioral factors, 8-9, 78
Vaginal cytology, 36-37, 39, 81
Vasodilators, 91
Venipuncture, 7-9, 70, 78
Ventricular septal defects (VSDs), 82
Vermin, 13, 15-16, 18-19
Veterinary care
aging, 80-81
cardiovascular diseases, 83, 85-86
Ehlers-Danlos syndrome, 92
emergency, 29
endocrinologic diseases, 97
health certificates, 31-32
hematologic diseases, 99-101
immunologic diseases, 104-107
infectious-disease control, 53-57
lysosomal storage diseases, 109-110
muscular dystrophy, 111-112
neurologic diseases, 113
ophthalmologic diseases, 115
orthopedic diseases, 116-117
pain and distress, 63-68
parasitic-disease control, 57-63
parturition, 40
procurement, 52-53
record-keeping, 28
surgery and postsurgical care, 68-70
see also Euthanasia; Surgery; Vaccines and vaccination; *and specific diseases*
Viral diseases, *see* Infectious diseases; *and specific diseases*
von Willebrand's disease, 97-98, 101

"*W*alking dandruff" (*Cheyletiella yasguri*), 58
Warts, 54
Washing facilities, 15
Water and watering devices, 26-27, 29-31, 53, 78, 80, 95, 99, 109, 111, 113
Weaning, 42-43, 46
Well-being
definition and measurement, 21
exercise, 22
Whipworm (*Trichuris vulpis*), 59, 61

X rays, 118
xmd dogs, 110-112

*Z*oonoses, 8-9, 37, 55, 62